AF687204

Tokenization and Digital Assets

Beyond the Hype of Blockchain in Finance

Joseph Trek

Published by NobleTrex Press

For permissions and other inquiries, write to:

P.O. Box 3132, Framingham, MA 01701, USA

Contents

Introduction **1**

1 Financial Foundations for the Digital Age **7**

1.1 The Bedrock of Wealth: Budgeting and Liquidity . . . 7

1.2 The Mathematics of Money: Interest, Inflation, and Time 18

1.3 Investing 101: Asset Classes and Risk 28

1.4 The Psychology of Money: Avoiding Behavioral Traps 36

2 The Technology Engine: How Tokenization Works **47**

2.1 The Digital Ledger: Consensus and Trust 47

2.2 Programmable Money: Smart Contracts 56

2.3 The Tokenization Lifecycle: From Rights to Tokens . . 64

2.4 Scale and Privacy: Building for the Real World 72

3 The Digital Asset Taxonomy **81**

3.1 Digital Cash: Payment Tokens and Stablecoins 81

3.2 Bridging the Gap: Security Tokens and RWAs 89

3.3 Access and Governance: Utility Tokens 95

3.4 Unique Digital Property: NFTs 101

4 Market Infrastructure and Operations **109**

4.1 Self-custody and key management 109

4.2 Institutional safekeeping: Custody models 121

4.3 Trading venues: CEX, DEX, and Liquidity 131

4.4 Moving value: Settlement and bridges 141

5 Guardrails: Regulation, Compliance, and Risk 151

5.1 The Regulatory Perimeter: Is It a Stock, a Commodity, or Just Magic Internet Money? 151

5.2 Compliance and Financial Crime: Why You Can't Stay Anonymous Forever 160

5.3 The Taxman Cometh: The Nightmare of Record Keeping 166

5.4 Operational Risk and Fraud: Don't Get Rugged 173

6 Strategy: Building a Digital Asset Portfolio 181

6.1 Valuation: Pricing the Priceless 181

6.2 The DYOR (Do Your Own Research) Field Guide . . . 193

6.3 Sizing the Bet: Allocation and Risk Management . . . 205

6.4 Making Assets Work: Yield and DeFi 214

7 The Future: Enterprise and Capital Markets 225

7.1 Show Me the Money: The Business Case for Tokenization 225

7.2 Rewiring Wall Street: Bonds, Funds, and Private Markets 236

7.3 ID Cards for Robots: Identity and Compliance-by-Design 246

7.4 Speaking the Same Language: Standardization and Coordination 255

Introduction

Here is a question that might make you uncomfortable: When you check your banking app and see a balance of $5,000, where is that money?

I don't mean "which branch is the cash sitting in," because we both know there is no vault with your name on it. I mean, what *is* it?

The reality is that your bank balance is nothing more than a row in a database controlled by a private company. When you "send" money to a friend, you aren't actually moving anything. You are sending a message to your bank, asking them to update their spreadsheet, while they send a message to your friend's bank, asking them to update theirs. If it's a weekend, or a holiday, or if the system is down, or if the fraud algorithm gets a little too trigger-happy, that money doesn't move.

For all the sleek interfaces and instant notifications of modern fintech, the plumbing of our financial system is surprisingly archaic. We are sending digital emails, but our money is still traveling by stagecoach. It settles in "T+2" days. It moves through a labyrinth of clearinghouses, correspondent banks, and custodians, each taking a tiny toll and adding a tiny delay.

Now, imagine if we could strip away those layers. Imagine if value could move as fast as information. Imagine if a share of stock, a deed to a house, or a dollar bill behaved like a text message-instant, global, and programmable, without needing a middleman to say "I approve."

This is the promise of tokenization.

It is not just about Bitcoin, and it is certainly not about buying pictures of bored apes for the price of a luxury sedan. It is about upgrading the operating system of capitalism.

You are reading this book because you suspect something important is happening, but the signal is drowning in the noise. You are right on both counts. We are standing on the precipice of the biggest shift in market structure since the invention of double-entry bookkeeping. But to understand it, we have to wade through a swamp of hype, scams, and technical jargon that makes quantum physics look like light reading.

That is where we come in.

The Great Unbundling of Finance

For the last century, financial trust has been institutional. We trust the bank to hold our money. We trust the broker to hold our stocks. We trust the title company to verify we own our home. These institutions are the gatekeepers of truth.

Blockchain technology-the engine behind tokenization-shifts trust from *institutions* to *mathematics*.

When we tokenize an asset, we create a digital container for it on a shared, immutable ledger. This container can represent anything. It can be a currency (like a stablecoin), a piece of art (an NFT), a vote in a community (a governance token), or a fraction of a commercial building in downtown Manhattan (a real-world asset).

Because these tokens live on a blockchain, they are "programmable." We can write code-smart contracts-that tells the money what to do.

Think about a vending machine. It is a primitive automated contract: *if* money is inserted, *then* release the soda. You don't need a lawyer or a cashier to oversee the transaction. The machine enforces the rule.

Tokenization allows us to build "vending machines" for complex finance. We can program a bond to automatically pay interest to whoever holds it in their digital wallet at 5:00 PM on Friday. We can program a royalty payment that splits instantly among five songwriters the second a track is purchased. We can program a savings account that automatically routes your deposit to the lending protocol offering the highest yield that second.

This creates a level of efficiency and liquidity that the traditional world simply cannot match. It turns money from a static noun into an active verb.

Beyond the Casino

I know what you are thinking. "If this technology is so great, why is the news filled with stories of exchanges collapsing, teenagers losing their savings on meme coins, and regulators handing out subpoenas like candy?"

That is a fair question. In fact, it is the *most* important question.

The crypto industry has a dual personality. There is the "Casino"-the speculative frenzy, the gambling, and the greed. And then there is the "Laboratory"-the engineers and financial architects building the rails for the next generation of global markets.

Most of what you see on Twitter or CNBC is the Casino. It's loud, it's volatile, and for the unprepared, it is dangerous. But if you dismiss the entire industry because of the Casino, you miss the breakthrough happening in the Laboratory.

Major investment banks aren't building trading desks for dog coins; they are tokenizing money market funds. Governments aren't experimenting with digital currency to facilitate drug deals; they are trying to modernize their settlement infrastructure. The technology is graduating from the fringes to the mainstream, and as it does, the "Wild West" era is slowly ending.

This book is designed to be your map through that transition. We are going to ignore the "get rich quick" schemes and focus on the "build wealth sustainably" mechanisms. We aren't here to hype you up; we are here to smarten you up.

How We Will Tackle This

We are going to take a slightly different approach than most crypto books. Usually, authors dive straight into the deep end of cryptography and hash functions. We aren't doing that.

You cannot build a skyscraper on a swamp. Before we touch a single digital asset, we have to ensure your financial foundation is solid. The first part of our time together will be a refresher on the timeless laws of money-liquidity, compounding, and risk management. If you don't have an emergency fund, you have no business owning Bitcoin. If you don't understand how inflation eats your purchasing power, you won't appreciate why scarce digital assets have value. We will ground ourselves in the classics before we launch into the avant-garde.

Once we have our footing, we will open up the engine. We need to demystify how this stuff actually works-not so you can code a blockchain, but so you can understand the risks. You need to know *why* a blockchain is secure, what a "smart contract" can and cannot do, and the difference between a decentralized network and a database that's just pretending to be one.

From there, we will explore the taxonomy of this new world. We will look at "Digital Cash" and stablecoins, which are rapidly becoming the killer app for payments. We will examine "Security Tokens," which is where the multi-trillion-dollar world of stocks and bonds is migrating. We will look at "Utility Tokens" and NFTs, moving past the cartoon jpegs to understand digital identity and property rights.

But theory is useless without practice. We will roll up our sleeves and talk about the plumbing. How do you actually store these things safely? The concept of "self-custody"-holding your own private keys-is powerful, but it is also unforgiving. If you lose your password at a bank, you call customer service. If you lose your private key, your money is gone forever. We will teach you the operational hygiene required to be your own bank without bankrupting yourself.

We will also tackle the elephant in the room: Regulation. The rules are changing fast. We will look at the legal landscape, the tax implications that catch people off guard, and how to spot the red flags of fraud. If a project promises you 20

Finally, we will look at how to fit this into a portfolio. How do you value an asset that has no cash flow? How much exposure is too much? We will apply institutional-grade due diligence frameworks to this chaotic market. And we will conclude by looking forward, beyond the retail hype, to how major enterprises are using this tech to slash costs and open new markets.

The Mental Shift

As we begin this exploration, I want to ask one thing of you: keep an open mind, but keep your skepticism handy.

We are discussing a technology that challenges deep-seated beliefs about value. We are taught that money is something the government prints. We are taught that banks are necessary for safety. We are taught that art is physical. Tokenization challenges all of these axioms.

It is easy to look at the volatility of digital assets and dismiss the whole thing as a bubble. It is equally easy to drink the Kool-Aid and believe that this technology solves every problem known to man. The truth, as always, is somewhere in the middle.

Blockchain is a tool. It is a hammer. You can use a hammer to build a house, or you can use it to smash your thumb. The difference lies in the skill of the user.

By the end of this book, you won't just be an observer of the digital asset revolution. You will be a literate participant. You will understand the difference between price and value. You will know how to separate a revolutionary protocol from a Ponzi scheme. And you will be equipped to make decisions that serve your financial goals, rather than your fear of missing out.

The world of finance is changing. The paper era is ending; the programmable era is beginning. It's time to learn the language.

Let's get started.

Chapter 1

Financial Foundations for the Digital Age

Before we launch into the futuristic world of tokenization and smart contracts, we need to check our parachutes. Many investors rush into digital assets treating them like lottery tickets, only to find that the ancient laws of finance-liquidity, inflation, and risk-still apply in the digital realm. This chapter isn't just a refresher; it's your survival kit. We'll translate dusty economic concepts into modern survival skills, ensuring you have the solid financial footing required to withstand the high-speed volatility of the blockchain era without losing your shirt-or your sleep.

1.1 The Bedrock of Wealth: Budgeting and Liquidity

Every bull market creates the same optical illusion: it looks like the asset is doing the hard work. Prices go up, screenshots get posted, and suddenly *discipline* feels optional—like an upgrade you can buy later, after the "real" money is made. Then real life shows up on schedule: a rent increase, a car repair, a layoff, a medical bill, a family emergency, a tax payment you underestimated. And if your finances are tight, the market doesn't need to scam you to hurt you. It only needs to dip at the wrong time.

That's the quiet villain of personal finance: the *forced sale*. Not the dramatic "I panic-sold everything" confession—more mundane than that. You sell because you have to. You sell at a bad price because the landlord doesn't accept "I'll pay you after the rebound" and the utility company doesn't offer an "HODL" discount. The forced sale is how perfectly good long-term investing plans get kneecapped by short-term cash needs.

The cure isn't a hotter tip, a more diversified watchlist, or an "all-weather" token portfolio. The cure is boring on purpose: predictable cash flow, and liquidity you can touch when the timing is terrible.

Financial Hierarchy of Needs

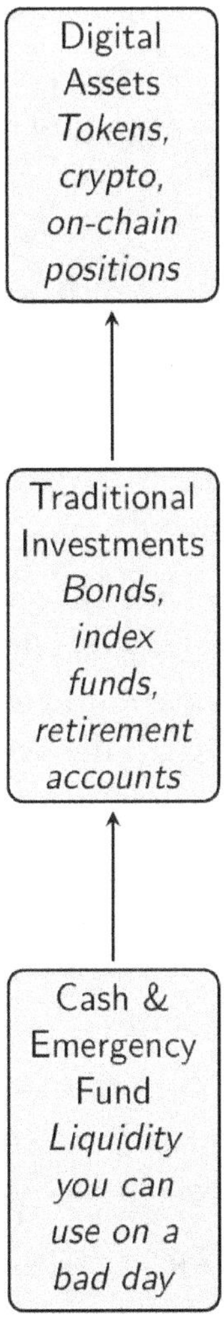

The pyramid is a reminder that money has *physics*. A structure can be tall, elegant, and ambitious, but only if the bottom is wide and solid. Digital assets—whether that's a major cryptocurrency, a tokenized fund, or a yield strategy that looks like free lunch in a glossy interface—sit at the top because they tend to be the least forgiving when timing goes against you. Even if the long-run story is compelling, the short-run price can be rude.

Liquidity is what turns a scary week into a merely annoying one. Without liquidity, volatility doesn't just move numbers on a screen—it reaches into your real life.

The financial runway: how long you can stay airborne

Budgeting has a branding problem. The word evokes spreadsheets, guilt, and the spiritual vibe of a dentist's waiting room. But the actual purpose of budgeting is closer to aviation: it tells you how much runway you have before gravity wins.

Your *burn rate* is the amount of cash your life consumes in a month once income and expenses have had their argument. If your burn rate is negative, you're saving; if it's positive, you're slowly setting your future on fire. Most people don't need a perfect budget. They need an honest one—especially if they're going to own anything that can drop 30% in a week and call it "normal."

A useful starting point is a cash-flow snapshot. Not aspirational. Not "what I usually spend." What happened. The past three months are a better therapist than your intentions.

Once you can see the numbers, the next step is almost embarrassingly mechanical: build slack. Slack is what separates investors from gamblers. Slack is the ability to wait.

Without slack, any investing plan—especially one involving tokens—quietly depends on everything going right in your personal life at the same time. It assumes your paycheck is stable, your health is stable, your car is stable, your landlord is stable, your taxes are predictable, and the market is cooperative. That's not a plan. That's a wish.

Monthly Cash Flow	*Month 1*	*Month 2*	*Month 3*
Take-home income (after tax)	$	$	$
Fixed essentials (rent/mortgage, utilities, insurance)	$	$	$
Food (groceries + dining)	$	$	$
Transportation (gas/transit, repairs, payments)	$	$	$
Debt minimums (credit cards, loans)	$	$	$
Health (premiums, copays, prescriptions)	$	$	$
Subscriptions & misc.	$	$	$
Net cash flow (income minus spending)	$	$	$

Table 1.1: A simple three-month reality check. Precision beats complexity; "rough but true" is better than "detailed but fictional."

Liquidity risk: the market doesn't care about your calendar

Liquidity risk is the danger that you'll need cash precisely when your assets are down, hard to sell, or both. It's not abstract. It's Tuesday at 4:47 p.m. when a payment is due and your portfolio is red.

Traditional finance people talk about liquidity like it's a property of markets—bid-ask spreads, trading volume, settlement cycles. For households, liquidity is more personal: *Can I pay the bill without detonating my future?*

Here's the part that trips up smart people: an asset can be valuable and still be a terrible emergency tool.

A stock index fund is a great long-term building block, but selling it in a panic is the opposite of why you bought it.

A rental property might be worth a lot, but you can't sell half a kitchen to cover a $1,200 car repair.

A stablecoin might be designed to hold a stable value, but it still lives inside a system with its own failure modes: exchange outages, withdrawal delays, banking cutoffs, smart-contract risk, network congestion, frozen accounts, changing terms, and the evergreen possibility that the "stable" part is truer in marketing than in crisis.

Even *cash* has nuance. Money in a bank account is generally far more liquid than money in an investment account, but "available" can still depend on timing, transfer methods, and holds. Liquidity is a chain, and it only takes one weak link—an app that won't load, a transfer that takes days, a card that gets flagged, an account that gets locked—to turn a small problem into a large one.

So the emergency fund isn't a vibe. It's a piece of risk management infrastructure.

The emergency fund: buying the right to do nothing

An emergency fund buys you something priceless in volatile markets: the right to do nothing.

When prices drop, the best investors don't display superhuman courage. They display *optional* behavior. They can wait. They can rebalance calmly. They can keep contributing on schedule. They're not staring at a red chart while doing mental math about rent.

That's what the emergency fund is for: it prevents your life from becoming correlated with your portfolio.

A practical emergency fund also creates a psychological firewall. When you know your next few months are covered, you're less vulnerable to the two worst sentences in investing:

> "I'll just sell a little to get through this."
>
> "I'll make it back quickly."

Those sentences are how temporary problems become permanent damage.

How big should the fund be? Rules of thumb range widely because lives range widely. Someone with stable income, strong insurance, and family support has a different risk profile than someone freelancing with variable pay and high fixed costs. The more fragile your income, the more valuable your liquidity. The more responsibilities you carry—kids, dependents, aging parents—the more expensive surprises become.

The better question is: *How many months of essential expenses does your runway cover?* Essential expenses are the boring non-negotiables: housing, utilities, basic food, insurance, minimum debt payments, transportation to keep earning. Not vacations. Not "my usual lifestyle." The number that keeps the lights on.

If you don't have an emergency fund yet, the first milestone isn't perfection. It's traction. A small buffer—$500, $1,000, one month of essentials—can stop the bleeding by preventing new credit card debt when life inevitably gets loud. Then the fund grows from "bandage" to "shock absorber."

The key is that the emergency fund must be *liquid in practice*, not liquid in theory. That typically means a checking or savings account at a boring institution with boring protections and boring customer service hours. In the hierarchy, "boring" is a feature.

Why token markets punish thin liquidity

Digital assets add a particular kind of cruelty to the forced-sale problem: the speed of drawdowns. Stocks can drop quickly, but many tokens can gap down in ways that feel like an elevator cable snapping—because liquidity can vanish when everyone tries to exit at once.

On top of that, the infrastructure layer matters. In traditional markets, the plumbing is old, regulated, and designed around institutions. In token markets, the plumbing is newer, more fragmented, and sometimes stress-tested in public. During periods of extreme volatility, you can face a stack of problems simultaneously:

- The price drops fast.
- The spread widens (selling costs more than you thought).
- The platform slows down or pauses withdrawals.

- The bank transfer you assumed was instant is not instant.

None of those require bad luck individually. They show up together because stress is correlated. Markets panic as a crowd, and systems fail as a crowd.

An emergency fund is how you keep the freedom to choose: hold, buy slowly, or simply ignore the noise. Without it, you might be forced to convert the most volatile part of your financial life into rent money at the exact moment it offers the worst exchange rate.

A budget that works: simple, honest, and slightly ruthless

A good budget isn't a document you obey. It's a feedback loop you trust.

Start with the principle that feels almost too obvious to write down: the budget has to match the life you actually live. If your plan requires you to become a different person—someone who never forgets to cook, never impulse-buys, never takes an Uber, never gives gifts, never has an inconvenient craving—it's not a budget. It's fan fiction.

Two habits make budgeting effective without turning it into a second job:

Name the fixed costs. Fixed costs are the heaviest items in your financial backpack. Rent or mortgage, car payments, insurance, subscriptions that have quietly become permanent. If cash flow feels tight, these are usually where the oxygen is disappearing. A $40 subscription won't save you from a $600 car payment you regret, but a $600 car payment will make every other decision feel like failure.

Separate "true expenses" from surprises. Many "emergencies" are not emergencies. They are predictable, just irregular: car maintenance, annual premiums, holiday travel, school costs, replacing a phone, medical deductibles, professional fees, taxes. The calendar isn't hiding; we're just not funding it. When you treat predictable costs as surprises, you end up borrowing from your future at the worst interest rate: stress.

This is where digital-asset enthusiasm can backfire. It's easy to fund "surprises" by selling whatever is up. That habit feels harmless when markets are kind, and it trains you to see your portfolio as a checking account. Then a down market arrives and the habit doesn't stop—because the bills don't stop.

Cash management as system design (not self-control)

Most people don't fail at cash flow because they're lazy. They fail because the system is fragile. A fragile system demands constant willpower, and willpower is a finite resource—especially when the market is noisy, your job is demanding, and your phone is a slot machine.

A resilient system does a few unsexy things well:

It separates money by purpose. When emergency cash sits in the same place as spending cash, it quietly becomes spending cash. When it's separated—ideally in a distinct account—it becomes psychologically harder to raid. You're not "moving money around." You're breaking glass.

It automates the boring wins. If saving requires a monthly moment of virtue, it will eventually lose to a monthly moment of chaos. Automatic transfers turn saving into default behavior. Defaults are powerful because they keep working when you're tired.

It reduces the blast radius of mistakes. If one overspending weekend can cascade into missed bills, overdrafts, credit card debt, and then a forced sale of investments, the system is brittle. Buffer accounts and modest cushions keep a small mistake small.

There's a deeper benefit here that doesn't show up in budgeting apps: these systems reduce your exposure to financial narratives. When your essentials are covered and your emergency fund is real, you don't need a story to justify your next decision. You don't need to believe that a token is "about to pump" so you can pay for an expense. You don't need to romanticize risk. You can treat investing as what it is: a long-term tool, not a short-term rescue.

The temptation of the ``working emergency fund"

At some point, almost everyone asks a clever-sounding question: why let the emergency fund sit there doing nothing? Why not put it in something that earns yield—money market funds, short-term bond funds, stablecoin lending, on-chain liquidity pools, *something*—so it can "work"?

That instinct is understandable. It's also where people accidentally swap *liquidity* for *performance* without noticing the trade.

The emergency fund has one job: show up on time, at full value, with minimal friction. Anything that complicates those requirements is not an upgrade; it's a redesign. Maybe it works most days. But emergencies don't arrive on "most days." They arrive on the days when systems are stressed and correlations spike.

If you want yield, earn yield with money that is allowed to misbehave. Long-term investments can be volatile because they're not meant to pay next month's bills. Emergency money is meant to be unglamorous and obedient. The "return" you're getting is not interest—it's the elimination of forced sales, late fees, high-interest debt, and the panic decisions that destroy compounding before it even starts.

A helpful mental model is to treat the emergency fund as an insurance premium you pay to the future version of yourself. Insurance premiums feel annoying right up until the day they feel brilliant.

A practical runway plan for would-be digital-asset investors

If you're eager to buy digital assets, the goal isn't to shame that impulse or smother it under caution tape. The goal is to finance it properly.

Start by making the runway visible: calculate essential monthly expenses, then compute how many months you could cover with cash that is truly liquid. If that number is uncomfortably low, the market is not your problem yet. The market will still be there after you fix the foundation. What won't be there, if you skip the foundation, is your ability to hold through volatility without negotiating with your landlord.

Then build the emergency fund in layers:

Layer 1: Immediate buffer. A small amount that prevents everyday mishaps from turning into debt: a deductible, a car repair, a plane ticket to deal with a family emergency.

Layer 2: Core runway. Months of essential expenses that protect you from income disruption. If you've ever felt your stomach drop at the words "restructuring" or "hours are being cut," you already understand the value.

Layer 3: Stability extras. The unsexy but life-changing category: planned irregular expenses, and a little margin for the fact that emergencies are rarely tidy.

Only after those layers exist does it make sense to allocate serious money to high-volatility assets. Not because tokens are uniquely evil, but because volatility plus thin liquidity is a trap. A good emergency fund breaks the trap open.

The paradox is that this "boring" work makes digital-asset investing more rational, not less. When you've bought yourself time, you stop needing every trade to be a winner. You stop needing the next narrative to be true. You can take smaller positions, hold longer, rebalance more calmly, and avoid the classic cycle of buying excitement and selling relief.

Wealth rarely collapses from one dramatic mistake. It erodes from preventable situations: paying a penalty because cash was tight, carrying high-interest debt because the buffer was missing, selling investments at the bottom because the bills were due. Budgeting and liquidity are the defenses that keep your long-term capital from being conscripted into short-term emergencies.

A strong foundation doesn't make you richer overnight. It makes you *harder to break*. And in markets that specialize in testing patience, being hard to break is a superpower.

1.2 The Mathematics of Money: Interest, Inflation, and Time

Money has a reputation for being concrete: numbers in an app, balances on a statement, a price tag on a shelf. In reality, money is a moving target. It changes shape as time passes, not because the digits morph, but because the world around those digits refuses to sit still. A dollar today is not the same tool it was ten years ago, and it won't buy the same slice of life ten years from now. Ignoring that is how people end up "saving" in a way that still makes them poorer.

Three forces do most of the work—quietly, constantly, and without asking permission: interest, inflation, and time. Put them together and you get the Time Value of Money (TVM), a concept that sounds like a textbook and behaves like gravity. TVM is simply the idea that a dollar now is worth more than a dollar later because money can earn a return, and because prices usually rise. In other words: time is not a neutral backdrop. Time is an active ingredient.

This matters in digital assets because the marketing is often allergic to math. You'll hear *APR* and *APY* thrown around like confetti, you'll see "stable" yields displayed with the confidence of a weather forecast, and you'll be tempted to treat any positive number as a win. But a return that looks good in nominal terms can be mediocre—or even negative—once you account for inflation and risk. The brain loves nominal returns because they're easy to brag about. The future version of you cares about real returns, because real returns buy groceries.

Nominal vs. real: the difference between ``more dollars'' and ``more life''

Start with a distinction that clears up a lot of confusion:

Nominal means "as stated." If a savings account pays 4% or a token strategy advertises 8%, that number is nominal. It is expressed in dollars (or tokens) without adjusting for how the purchasing power of those dollars changes.

Real means "after inflation." Real return is the result that matters for your actual standard of living.

The basic approximation is refreshingly blunt:

$$\textit{Real Return} \approx \textit{Nominal Return} - \textit{Inflation}$$

This isn't a fancy formula. It's a reality check.

If you earn 5% in a year when inflation is 3%, your real return is roughly 2%. Your account balance rises faster than prices, so you can buy more stuff next year than you can this year. If you earn 5% when inflation is 6%, your real return is roughly -1%. Your balance might look bigger, but your money buys less. You have more digits and less purchasing power—the financial version of running on a treadmill that quietly speeds up.

This is why "I'm earning yield" is not the same sentence as "I'm getting ahead."

Inflation: the silent thief with perfect manners

Inflation rarely feels like a single event. It's not a dramatic headline most days; it's a slow accumulation of small indignities. The coffee price nudges up. The grocery bill stops making sense. Insurance renewals arrive with that cheerful, infuriating optimism: "Your premium has been updated."

In finance, inflation is the general rise in prices over time. In human terms, inflation is the shrinking of your money's purchasing power. It is a tax you pay without writing a check.

A helpful way to think about it: cash has a hidden interest rate, and it's often negative. If your dollars earn nothing, and the cost of living rises, your cash is being diluted. It's the safest-looking asset that can quietly become risky over long horizons.

Inflation is also why "I'll just hold cash until I feel ready" can become a long-term trap. Waiting feels prudent. But if waiting lasts years, and inflation keeps nibbling, you can lose meaningful purchasing power while feeling like you're doing nothing wrong. The thief is polite. It doesn't kick the door in. It simply rearranges the value of everything while you're living your life.

Of course, inflation doesn't move in a straight line. Some years it's mild; some years it makes people re-price their entire relationship with money. That variability is one reason it's dangerous to anchor on a single number. The concept is what matters: if you're measuring progress in nominal dollars alone, you're measuring with a ruler that stretches.

Compounding: the snowball that doesn't look impressive at first

If inflation is a silent thief, compounding is a patient builder. It's the reason long-term investors sound boring and end up wealthy anyway.

Compound interest means you earn a return not only on your original money (your *principal*), but also on the returns you earned previously. The snowball metaphor gets used so often it risks becoming wallpaper, but it's accurate for a specific reason: the growth accelerates.

Early on, compounding looks underwhelming. A few percent on a modest balance is not life-changing. This is why people chase shortcuts. They want a graph that looks exciting in month one.

But compounding has an unfair advantage: it keeps going while you sleep, and it gets more powerful as the base gets larger. The first ten years are the awkward phase. The later years are where the curve starts acting like it has opinions.

And compounding isn't just about the return rate. It's also about *time* and *contributions*. Someone who invests modestly but consistently can outrun someone who "waits for the perfect moment" and then invests in a hurry. Markets reward persistence more reliably than they reward timing.

This is an important antidote to digital-asset culture, which often treats time as an obstacle rather than a weapon. "Getting rich quick" is seductive because it feels like skipping the line. Compounding is less dramatic. It's also far more likely to work.

A picture worth twenty years: cash erodes, compounding climbs

The cleanest way to feel the difference between nominal balances and real purchasing power is to put two lines on a chart and let time do the arguing.

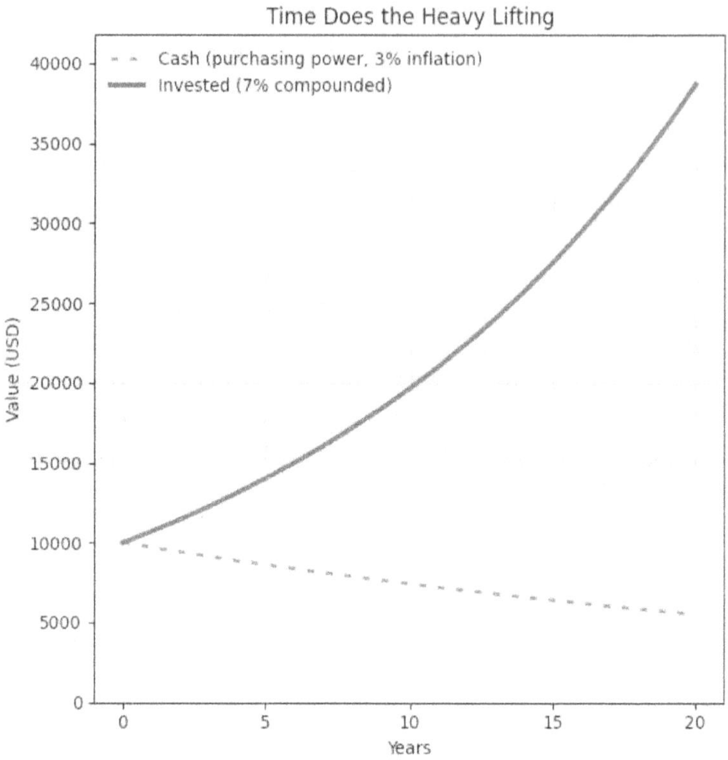

The chart hides a small philosophical punch. The "cash" line isn't showing your bank account shrinking. It's showing your *ability to buy things* shrinking. Your statement can look stable while your life gets more expensive. That's inflation's trick: it doesn't need to touch your balance to take value from you.

The compounding line, meanwhile, starts out politely. Then it bends upward, and by the later years it becomes less about the annual rate and more about the fact that the base is bigger. Compounding turns time into leverage.

Neither line is a promise, of course. Markets don't guarantee 7%. Inflation doesn't hold at 3%. The point is the direction of the forces. If you ignore inflation, you can accidentally accept a long-term loss while feeling cautious. If you ignore compounding, you can underestimate how much "boring consistency" can outperform flashy tactics.

APR, APY, and the seduction of small print

Digital-asset platforms love yield displays because yield is the easiest part to sell. But yield comes with vocabulary that's designed to sound familiar while hiding sharp edges.

APR (Annual Percentage Rate) is usually the stated annual rate, not accounting for compounding within the year.

APY (Annual Percentage Yield) includes the effect of compounding. If interest is paid and reinvested frequently—daily, weekly, per block—APY will be higher than APR for the same base rate.

Here's the problem: platforms can emphasize whichever number looks prettier, and you can't evaluate either number without understanding *what you're being paid in* and *what risks you're taking*.

A 12% APY paid in a token that can drop 50% is not the same as 12% paid in dollars. A 6% yield that depends on continuous access to a smart contract is not the same as 6% from a Treasury instrument that has the U.S. government's credit behind it. And a "stablecoin" yield might be stable until the day it isn't—because the stability is a design goal, not a law of nature.

The math of money is allergic to sales pitches. It forces you to ask the questions that marketing would prefer you skip.

Real returns in the wild: ``5% on a stablecoin"

Consider a common claim: "I'm earning 5% yield on a stablecoin."

If inflation is 3%, the quick-and-dirty real return is about 2%—before fees, taxes, and friction. That might still be attractive for money you're comfortable putting at risk *in that way*. But it's not the windfall it sounds like in casual conversation.

Now add a layer that people often omit: risk is part of the price. Traditional finance has a harsh but useful rule—higher potential return usually requires higher risk. If you see a yield that looks generous relative to low-risk alternatives, the correct response is not gratitude. It's curiosity.

Where is the yield coming from? Is it coming from lending to leveraged traders? From liquidity incentives funded by token issuance? From a strategy that works until volatility spikes? From a counterparty who can say "withdrawals are temporarily paused" on the day you need liquidity most?

If the answer is "I'm not sure," then the yield is not just a number. It's a mystery you're betting your money on.

The deeper point is not that yield is bad. The point is that *real return* is a three-part sentence:

> What did I earn, after inflation, after costs, for the risk I took?

If you don't finish the sentence, your brain fills in the ending with optimism.

Discounting: why a future dollar is worth less (even if you trust the payer)

TVM also works in reverse. Not just "money grows over time," but "future money is worth less today." This is called *discounting*. It's how professionals decide what a stream of future payments is worth right now.

You don't need to memorize the formula to benefit from the concept. Discounting answers questions like:

Should I take $10,000 today or $10,000 in five years?

If you can earn a return in the meantime, and if inflation erodes purchasing power, the later payment is worth less. Even when the dollar amount is identical, the economic value is not.

Discounting is also a quiet weapon against hype. When a project promises future utility, future adoption, future revenue share, future token burns, future everything—discounting is the tool that asks: *How much is that future worth today, given uncertainty and time?*

Many token narratives are basically long strings of future-tense verbs. Discounting forces them to pay rent in present-tense math.

Inflation hedges and the idea of ``keeping up"

If inflation is the problem, one obvious goal is simply to keep up—to avoid losing purchasing power. Some instruments are designed with that in mind.

Treasury Inflation-Protected Securities (TIPS), for example, adjust their principal based on inflation measures (commonly tied to CPI). The concept is elegant: if inflation rises, the principal rises, and interest payments change because they're calculated on the adjusted principal. At maturity, there's typically a floor so you don't receive less than the original principal even if there's deflation. TIPS won't make you rich. That's not their job. Their job is to be a more honest contract with reality when "cash but safe" isn't safe in purchasing-power terms.

Even if you never buy TIPS, the existence of such instruments teaches a useful lesson: you can separate "return" into different jobs. Some returns are meant to build wealth. Some are meant to preserve purchasing power. Confusing those jobs is how people end up taking speculative risks with money that was supposed to stay boring.

Digital assets often blur these categories on purpose. A token can claim to be a currency, a growth asset, a savings instrument, and an inflation hedge—sometimes in the same paragraph. TVM is the discipline of asking: *Which job is this money doing, and what happens if it fails?*

The tax and fee reality: your portfolio doesn't compound at the headline rate

Compounding is powerful, but it's also fragile. It depends on what you get to keep.

Fees act like a slow leak. A 1% annual fee doesn't sound like much until you realize it gets paid every year, on a growing balance, forever. Over long periods, small percentage differences compound into large outcome differences.

Taxes matter too, because they can interrupt compounding. If a strategy generates frequent taxable events, your "return" is partly a return you must share on a schedule. Digital-asset strategies can be especially messy here: staking rewards, airdrops, liquidity incentives, and frequent trades can create complicated tax profiles depending on jurisdiction and specifics. The key point is conceptual: the number you see on a dashboard is rarely the number that compounds in your real life.

The headline rate is what marketing wants you to remember. The after-tax, after-fee, after-inflation return is what your future wants you to calculate.

Time horizon: the hidden variable that changes the whole equation

The same investment can be "too risky" and "perfectly reasonable" depending on when you need the money.

Time horizon is simply how long you can leave the money alone. If you need the money next year, volatility is not an abstract concept—it's a direct threat to your plan. If you don't need the money for twenty years, volatility becomes something closer to background noise, assuming you can stay invested and you're diversified appropriately.

This is where compounding and inflation become a pair of competing incentives. Inflation pushes you to invest so your money doesn't rot in place. Volatility pushes you to hold safer assets when your time horizon is short. The right answer isn't a slogan. It's alignment: match the risk of the asset to the job the money is doing and the time you can give it.

Digital assets, by their nature, tend to be poor tools for short horizons. Not because the technology is doomed, but because prices can swing wildly for reasons that have nothing to do with your personal calendar. If you might need the money for a move, a wedding, a tuition bill, or a down payment, volatility is not "exciting." It's sabotage.

A mental model that survives hype

A durable investing mindset is less about predictions and more about arithmetic.

First, measure success in real terms. Ask what your returns mean after inflation, not just in nominal dollars.

Second, treat compounding like a machine that you protect. Anything that forces you to stop and start—panic selling, constant strategy switching, excessive fees, avoidable taxes—damages the machine.

Third, understand that yield is not magic. It is compensation for something: time, illiquidity, risk, leverage, complexity, or someone else's demand for capital. If you can't name what you're being compensated for, you don't understand the product—you're just enjoying the number.

None of this requires advanced math. It requires refusing to be hypnotized by nominal figures and glossy interfaces. The mathematics of money isn't there to impress you. It's there to keep you from being impressed by the wrong things.

1.3 Investing 101: Asset Classes and Risk

Risk has a PR team. It shows up wearing the costume you're most likely to trust: "opportunity," "innovation," "the future." In a boom, risk even feels like a personality trait—bold people take it, timid people don't, and the market rewards the brave. Then the cycle turns and you find out what risk really is: a bill that arrives without warning, payable in stress.

A cleaner definition is less dramatic and more useful: risk is the chance that reality won't cooperate with your plan. Sometimes that means losing money. Sometimes it means *not having money when you need it*. Sometimes it means watching a supposedly "safe" investment fail at the only job it had—like cash quietly failing to keep up with inflation, or a "stable" yield product failing exactly when everyone rushes for the exits.

Asset classes are simply the major categories of things people invest in. They're not just labels for finance nerds; they're different bundles of risk. Each one has its own way of disappointing you, and its own way of helping you. The trick is to stop asking, "What will perform best?" and start asking, "What problem is this money supposed to solve?"

At a high level, the menu doesn't change just because the wrapper becomes digital. Cash is still cash. Loans are still loans. Ownership is still ownership. Real assets still exist in the physical world. Tokenization can make these things easier to trade, easier to split into smaller pieces, and easier to program. It doesn't repeal the underlying math of risk and return.

The risk---return bargain (the one contract you can't cancel)

There's a bargain at the center of investing, and it's remarkably consistent across centuries and technologies: higher potential returns tend to require taking higher risk. People sometimes talk about this as if it's a moral lesson. It's not. It's a pricing mechanism.

If something is genuinely safe—meaning the likelihood of loss is low and the path is smooth—lots of people want it. That demand pushes the expected return down. If something is scary, uncertain, or volatile, fewer people want it. To convince buyers to show up anyway, the expected return has to be higher.

This is why "guaranteed high returns" should set off alarms. In finance, free lunches are rare, and the market is crowded with professionals trying to eat them first. When you see unusually attractive returns, the real question is what kind of risk you're being paid to hold—credit risk (someone might not pay), duration risk (interest rates might move), liquidity risk (you can't sell quickly), volatility risk (prices swing), or a more exotic digital risk (smart contract failure, platform failure, governance failure).

When people say "I have a high risk tolerance," they often mean "I can handle the feeling of volatility." That's only one dimension. Risk tolerance also includes practical tolerance: can your life withstand a drawdown without forcing you to sell?

The cast of characters: what each asset class really is

Asset classes are easier to remember when you strip away the tickers and focus on what you own.

Cash and cash equivalents are money you can access quickly, typically with low volatility. The return is usually low, and the hidden risk is inflation. Cash is not trying to make you rich. It's trying to keep you functional.

Bonds are loans. When you buy a bond, you're lending money to a government, a company, or another entity. In return, you receive interest and (usually) principal back at maturity. Bonds can be relatively stable compared to stocks, but they're not risk-free. Borrowers can default, and bond prices can fall when interest rates rise. Bonds are the financial equivalent of choosing steadier paychecks over a lottery ticket.

Stocks are ownership. When you buy a stock, you own a slice of a business and participate in its upside (and downside). Stocks have historically offered higher long-term returns than bonds, but they can be brutal in the short term. Stocks are where you go when you want growth and you're willing to tolerate drama.

Real estate is both a place and a balance sheet. It can generate income (rent) and appreciate over time, and it can also concentrate risk: leverage, vacancy, repairs, local economic downturns, and regulatory shifts. Real estate often feels stable because the price doesn't update every second on your phone. That calm is partly an illusion; the underlying value can still change.

Commodities are raw materials—oil, metals, agricultural goods. They can hedge certain risks, but they don't compound in the same way a productive business does. Commodities can be useful tools, but they don't love being the centerpiece of a long-term plan.

Digital assets are a broad family, not one thing. Some behave like commodities (a scarce, traded asset with no cash flows). Some are more like equity (claims on network fees or governance influence, though the legal reality can be messy). Some are more like software licenses, membership cards, or arcade tokens with a global trading venue attached. The volatility is often high, the narratives are loud, and the infrastructure is still evolving.

The point isn't to crown a winner. The point is to recognize that these categories exist because risks exist. If you don't choose your risks, you still get risks—you just get them accidentally.

Where digital assets fit: not the foundation, not the whole house

Digital assets tend to sit on the speculative end of the spectrum for a few reasons that have nothing to do with whether the technology is "good." Many tokens lack cash flows, so pricing can be driven heavily by expectations. Markets can be fragmented. Liquidity can change quickly. Regulatory frameworks are still catching up. And because trading is frictionless, sentiment can travel at the speed of a meme.

That doesn't mean digital assets are automatically foolish. It means they are rarely suitable as the money that must be stable—down payments, near-term tuition, emergency reserves, next year's rent. Digital assets can be appropriate for capital you can leave alone for a long time and that you can afford to see fluctuate.

The healthiest way to hold digital assets inside a portfolio is to give them a job description. Are they your "high-risk growth engine"? A small, asymmetric bet on a new financial infrastructure? A hedge against specific scenarios? A hobby allocation you can afford to lose? If you can't describe the job, you'll end up treating price action as the job—and that's how portfolios turn into slot machines.

Risk is a spectrum, not a binary

"Safe" and "risky" are lazy words. They flatten everything into a moral drama. Real investing decisions live on a spectrum. Seeing that spectrum—visually—helps keep your portfolio from becoming a referendum on your identity.

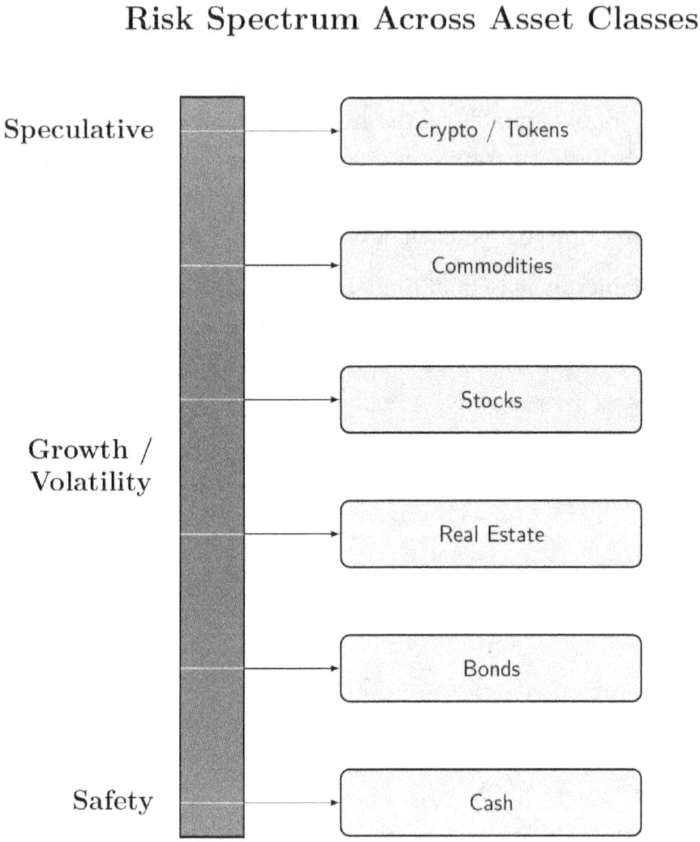

The slider is deliberately not a precise measuring tool. Real life is messier. A short-term Treasury bill and a long-term bond don't behave the same way. A diversified stock index fund and a single meme stock do not belong in the same emotional category. "Crypto" contains everything from relatively established networks to experiments that should come with a helmet.

Still, the spectrum is a useful antidote to two common mistakes: thinking that "safe" means "no risk," and thinking that "risky" means "always dumb." Safe assets have risks; risky assets can have a place. The art is in proportion, purpose, and time horizon.

Diversification: the only free lunch (and it's not a metaphor)

Diversification is finance's most overused good idea—and still the best one. It's simply the practice of spreading your money across different assets so that a single failure doesn't ruin you.

Why it works is the key: asset classes don't always move together. When one is having a bad year, another might be holding up, or even doing well. Diversification doesn't guarantee profits. It reduces the chance that your future depends on one fragile story.

This is why diversification is often called the only "free lunch" in finance. You can reduce risk without necessarily sacrificing expected return, because you're not just averaging outcomes—you're reducing the odds of a catastrophic one.

But diversification has a catch that people discover the hard way: owning *many things* is not the same as owning *different things*. Ten tokens can still be one bet if they all depend on the same market mood. Five tech stocks can still be one bet if they're all priced for the same growth narrative. Diversification is about different drivers of returns, not different logos.

Digital-asset markets can be particularly deceptive here because correlations often spike in stress. In calm periods, tokens can look independent. In panic, they can fall like dominoes that were always connected, just not obviously.

The most common portfolio mistake: concentrating without meaning to

People rarely set out to build a concentrated portfolio. Concentration usually happens by accident, through a series of decisions that each feel reasonable.

You buy the thing that's going up because it feels like confirmation. You avoid the thing that's lagging because it feels like dead weight. Over time, you end up with a portfolio that is essentially one big position in "whatever has been working lately." That portfolio feels great right up until the regime changes.

Traditional investors do this with hot sectors. Digital-asset investors do it with narratives. The names change—DeFi summer, NFT mania, Layer-2 rotations, AI tokens—but the behavior is old. It's performance-chasing in a new costume.

A diversified portfolio is an argument with your own attention. It says: I will not let the most exciting chart decide my financial future.

Rebalancing: disciplined discomfort, on purpose

Rebalancing is the most practical expression of diversification. It means periodically adjusting your portfolio back to target allocations when markets move.

If stocks outperform, they become a larger slice of your portfolio than you intended. Rebalancing trims them back. If bonds lag, rebalancing tops them up. In plain language, rebalancing forces you to do the thing that feels wrong in the moment: trim winners, add to losers.

That's not a gimmick. That's the point.

Without rebalancing, your portfolio tends to drift toward whatever has been rising. It becomes more aggressive at the top of the cycle—precisely when you're being paid the least for taking risk. With rebalancing, you're constantly restoring your original risk posture.

Digital assets make rebalancing emotionally harder because the swings are louder. When a token doubles, selling some feels like betrayal. When it collapses, buying some feels like catching a falling knife. Rebalancing doesn't require you to predict the future; it requires you to respect your own plan more than your current mood.

A small but important nuance: rebalancing is not automatically "buy low, sell high" in the meme sense. It's "keep risk where you intended it." The buy-low-sell-high effect is a side benefit of staying honest about risk.

Risk comes in flavors: volatility, liquidity, credit, and inflation

Volatility gets all the attention because it's noisy. But volatility is just one flavor of risk.

Liquidity risk is whether you can sell quickly at a fair price. A private investment can be valuable and still trap you at the wrong time. In tokens, liquidity can evaporate in moments, turning theoretical value into practical frustration.

Credit risk is whether someone pays you back. It matters in bonds, loans, and many yield products—traditional or on-chain. If you're earning interest, ask who is on the other side of the promise.

Inflation risk is the risk that "safe" assets fail to keep up with rising prices. Cash is the classic case: stable in nominal terms, unstable in real terms.

Operational and technological risk is more prominent in digital assets: custody mistakes, smart contract vulnerabilities, governance attacks, platform insolvencies, and failures in the surrounding infrastructure. These risks can be invisible during good times because everything works during good times.

Thinking in flavors matters because it stops you from treating risk like a single dial you turn up or down. You can sometimes reduce one risk and increase another. A long-term bond can reduce credit risk relative to a shaky borrower, but it can increase interest rate sensitivity. A yield-bearing token strategy can reduce inflation risk if it truly produces a return, but it can raise operational risk dramatically.

The goal isn't to eliminate risk. The goal is to take risks you are paid for, that you understand, and that your life can tolerate.

A practical framework: match assets to goals, not moods

The most stable investing plans are built from an unglamorous question: *When will I need this money, and what happens if it's down then?*

Money needed soon tends to demand stability and liquidity. Money meant for decades can take more volatility in exchange for expected growth. That's not conservatism—it's respect for time horizon.

Once you view investing as goal-matching, the asset class lineup becomes less confusing. Cash is for near-term needs and resilience. Bonds can dampen volatility and serve intermediate horizons. Stocks are the workhorse of long-term growth. Real estate can diversify and provide income, but it concentrates location and leverage risks. Commodities can hedge certain scenarios. Digital assets, for most people, belong as a deliberate slice of long-term risk capital—sized so that a drawdown is uncomfortable, not catastrophic.

Sizing is the part that turns philosophy into survival. A small allocation to speculative assets can be a healthy way to participate in innovation. A large allocation can turn your future into a hostage of market sentiment.

The market will always offer a new story. A portfolio built on asset classes, time horizon, diversification, and rebalancing doesn't need the story to be true. It needs the math to keep working—quietly, persistently, and in your favor.

1.4 The Psychology of Money: Avoiding Behavioral Traps

The most expensive part of investing isn't the management fee. It isn't even the occasional bad pick. It's the moments when your brain grabs the steering wheel and insists it can drive better than your plan.

That sounds insulting until you remember what your brain evolved to do. It was built for social survival, not capital allocation. It's brilliant at spotting threats, reading the mood of a crowd, and reacting quickly when something feels urgent. Markets—especially token markets—are essentially a machine that manufactures urgency. Price updates arrive like heartbeats. Group chats turn into trading floors. A line on a chart becomes a referendum on your intelligence. In that environment, "rational investor" is less a personality type and more a temporary state you rent.

Behavioral finance is the study of predictable mistakes people make with money. Predictable is the key word. These aren't random flaws sprinkled across humanity. They're patterns—common enough that researchers can name them, measure them, and watch them repeat. If you can recognize the patterns while they're happening, you can build systems that block the worst decisions before they feel irresistible.

FOMO: the fear of missing out (and the fear of being ordinary)

FOMO isn't just the fear that an asset will go up without you. It's the fear that other people will get a better future than you and you'll have to watch. It's status anxiety wearing a price chart as a mask.

In traditional markets, FOMO exists, but it often moves at the speed of quarterly reports and CNBC segments. In digital assets, it moves at the speed of screenshots. Someone posts a 10x. Someone else replies, "Still early." A third person adds rocket emojis, which is not analysis but is somehow persuasive. The social proof hits before your skepticism boots up.

FOMO has a signature behavior: it makes you buy after the big move, not before it. You don't feel compelled when the asset is boring. You feel compelled when the asset is already expensive—because that's when the crowd is loudest. Price becomes the marketing budget.

There's a cruel irony here. The more your purchase is driven by FOMO, the more likely you're buying into short-term euphoria. That's the exact moment when the risk is highest and the future returns—mathematically—are often lowest. FOMO turns the market into a game where the prize goes to whoever can stay calm while others are performing excitement.

A useful diagnostic question is embarrassingly simple: *If nobody could see this trade, would I still want it?* If the answer changes, you're not investing; you're auditioning.

Loss aversion: why losing $100 hurts more than winning $100 feels good

Loss aversion is one of the most documented quirks in behavioral finance. It means losses feel larger than gains of the same size. A $1,000 gain is pleasant; a $1,000 loss is personal. The brain treats losses like threats, and threats trigger action.

In markets, that emotional asymmetry creates two classic problems.

First, it can cause panic selling. When prices drop fast, you don't just see red numbers-you feel your future shrinking. Selling promises immediate relief. It turns uncertainty into certainty. Unfortunately, it can also lock in the loss right before a rebound, which is how "I couldn't take it anymore" becomes a permanent financial scar.

Second, loss aversion can cause the opposite: stubborn holding. People avoid realizing a loss because selling makes it *real*. As long as you hold, you can tell yourself it's temporary, a paper loss, a misunderstanding by the market. This is how portfolios accumulate "someday it'll come back" positions—assets held not because they fit the plan, but because selling would hurt.

The market doesn't care which version of loss aversion you prefer. It profits from both.

One of the sneakiest features of token markets is how they intensify loss aversion by keeping you constantly informed. If you check prices ten times a day, you experience ten emotional weather reports. Even when nothing changes materially, your nervous system gets trained to treat investing as an emergency. That training has consequences. It makes you more likely to act, and acting is often how investors underperform their own portfolios.

The disposition effect: selling winners, holding losers, and calling it ``discipline"

The disposition effect is a specific behavioral pattern observed in investors: we tend to sell assets that have gone up (to "lock in gains") and hold assets that have gone down (to "avoid locking in losses"). It sounds reasonable when phrased casually. It also creates a portfolio that, over time, can become a museum of regret.

Why does this happen? Because selling a winner gives you a clean hit of satisfaction. You were right. You did something smart. Your identity gets to cash a check. Selling a loser forces you to admit you were wrong, and the brain treats that admission as pain. So you delay it.

In hype-driven markets, the disposition effect can be extra destructive because of the way big winners behave. A small number of assets can drive a large portion of returns. If you constantly trim your winners early and keep nursing your losers, you're doing the financial equivalent of uprooting healthy plants and watering dead ones because it feels fair.

"Taking profit" is not automatically bad. It becomes bad when it's not connected to a coherent rule—when it's just emotion dressed up as prudence. The antidote isn't stubbornness. It's structure: position sizing, rebalancing rules, and exit criteria decided when you're calm.

Narrative fallacy: when a good story outruns the numbers

Humans are storytelling animals. We don't just want facts; we want meaning. We want the world to make sense in a way that fits into a sentence.

The narrative fallacy is what happens when we mistake a compelling story for a reliable forecast. We see a sequence of events and build a neat explanation that feels inevitable. Markets are especially vulnerable because stories are contagious, and because price movement itself gives stories credibility. A rising price makes the story feel true, even if nothing fundamental has changed.

Digital assets are a natural habitat for narrative fallacy. Many tokens don't have traditional valuation anchors like earnings. When valuation is hard, story becomes the substitute. "This network will be the settlement layer of the internet." "This token is digital gold." "This protocol will replace banks." These claims may be partially true, or eventually true, or never true—but story-based pricing can run far ahead of proof.

Narrative fallacy has a signature smell: it makes the future sound obvious. The more inevitable a story feels, the more suspicious you should be. The future is rarely obvious; it just becomes obvious in retrospect, after the winners are known and the dead projects have been swept away.

A brutal way to test a narrative is to translate it into a question with numbers. Not because numbers are perfect, but because they force specificity.

If a token is supposed to capture network value, *how* does value accrue—fees, burn mechanisms, revenue sharing, governance power, or pure speculation? If a project is "the future," *what* would adoption look like, and what would have to be true for that adoption to happen? If the answer is vague, you're not holding an investment thesis. You're holding a mood.

Recency bias: why the last three months feel like destiny

Recency bias is the tendency to overweight recent events when predicting the future. After a strong run, the brain starts treating "up" as normal. After a crash, it starts treating "down" as permanent.

Recency bias is the reason people buy at peaks and sell at bottoms while sincerely believing they are being prudent. At the peak, recent gains make risk feel low and opportunity feel urgent. At the bottom, recent losses make risk feel high and safety feel urgent. The same person can flip from "this is the future" to "this is a scam" based on a few weeks of price movement, even if nothing important has changed.

In token markets, recency bias is turbocharged by the cadence of information. In a single day you can absorb a week's worth of emotional stimuli: a regulatory rumor, a whale moving coins, a hack headline, an influencer thread, a sudden pump, a sudden dump, and a chart that looks like it was drawn by a seismograph. Your brain wants to protect you from that chaos by finding a simple rule. The simplest rule is: whatever just happened will keep happening.

That rule is almost always wrong.

Anchoring: the price you first saw becomes your reference point

Anchoring is when the first number you encounter becomes a mental reference point. In investing, the most common anchor is the price you bought at, or the all-time high you wish you'd sold at.

Anchors are sticky. If you bought at $100 and the asset drops to $60, your brain treats $100 as the "real" value and $60 as a temporary insult. You may hold even if new information suggests the investment no longer makes sense. Or you may buy more simply to get your average price down, because it feels like progress.

Anchoring also drives a subtle form of FOMO: "It used to be $200, so $120 is cheap." Maybe. Or maybe $200 was ridiculous.

Markets don't care about your anchor. The only question that matters is: given what you know now, and the alternatives available now, is this asset the best use of your money now?

That question is hard because it demands humility. Anchors let you avoid humility by pretending the past has authority.

Overconfidence: the bull market that happens inside your head

Overconfidence is what makes a good streak feel like skill and a bad streak feel like bad luck. It's also what makes people increase position sizes right when risk is rising—because success breeds certainty.

Digital markets are particularly good at manufacturing overconfidence because they can reward impulsive behavior for a while. When a market is ripping higher, almost any risk looks smart. You can buy something with a name that sounds like a joke and still win. The brain doesn't interpret that as "markets are exuberant." It interprets it as "I have a gift."

Then the market turns and that "gift" starts charging tuition.

Overconfidence also shows up as complexity addiction—the belief that a more complicated strategy must be smarter. Leverage, derivatives, yield loops, cross-chain bridges, exotic liquidity positions: these tools have legitimate uses, but in the hands of someone chasing excitement, they become a way to turn volatility into catastrophe. Complexity makes it easier to confuse activity with progress.

A calm portfolio often looks boring from the outside. That's not a failure. That's the point.

The cycle of market emotions (and why it keeps working on smart people)

If you want to understand why buying high and selling low happens at scale, don't imagine individual investors making isolated mistakes. Imagine a crowd moving through predictable emotional states.

It usually starts with curiosity: a new technology, a new asset, a new narrative. Then optimism: prices rise, early adopters look smart, the story starts to spread. Then excitement: participation becomes social. Then euphoria: risk is reframed as destiny. At that point, skepticism is treated as ignorance, and caution is treated as cowardice.

The top doesn't arrive with a bell. It arrives with a feeling: everyone you know suddenly has an opinion. Then the market dips, and instead of interpreting the dip as a normal feature of risk assets, the crowd interprets it as injustice. They buy the dip aggressively, confident the story will rescue them.

When the dip becomes a slide, emotions shift: anxiety, denial, bargaining, anger. Eventually you get capitulation—the moment when selling feels like the only way to stop the pain. That's often near the bottom, because maximum pain is what forces the most people out.

The cycle persists because it's rooted in human wiring. It doesn't require stupidity. It requires social brains in a volatile environment.

How to fight your brain without pretending you can ``out-discipline'' it

The goal is not to become a robot. The goal is to design a system that behaves like one when you can't.

The most effective defenses against behavioral traps are pre-commitments: rules you decide in advance, while your heart rate is normal. When volatility hits, you don't want to be negotiating with yourself. You want to be executing.

A few rules are especially powerful because they attack multiple biases at once.

- *Position sizing: make it impossible for one asset to ruin you.* If a single trade can wreck your sleep, it can wreck your portfolio.

The right size is the size you can hold through bad weeks without rationalizing desperate behavior. In digital assets, smaller position sizes are not timid; they are realistic.

- *Dollar-cost averaging (DCA): buy on a schedule, not on a mood.* DCA means investing a fixed amount at regular intervals. It reduces the temptation to "wait for the perfect dip" (which is often just recency bias wearing a strategy costume). It also reduces regret: you're not betting everything on one entry point.
- *Rebalancing: keep your risk where you intended it.* Rebalancing is a behavioral hack disguised as portfolio maintenance. It forces you to trim what's run up (countering FOMO and overconfidence) and add to what's lagging (countering panic and loss aversion), without requiring you to predict anything.
- *A sell rule: decide what would make you exit.* Many people have a buy thesis and no exit thesis. That's like getting into a plane without checking how the doors open. Your exit rule might be valuation-based, time-based, thesis-based ("if this key assumption breaks, I'm out"), or risk-based. The exact rule matters less than having one.
- *A "price-check budget": limit how often you look.* This sounds trivial until you try it. Attention is a lever. The more you watch, the more you feel. The more you feel, the more you act. If you must check prices constantly, you're not investing-you're feeding an anxiety engine.

These rules don't guarantee profit. They reduce self-sabotage, which is often the bigger problem.

A quick self-audit: are you investing, speculating, or coping?

A surprisingly large portion of trading is emotional coping disguised as strategy. People trade to feel in control, to escape boredom, to chase social belonging, to recover from regret, to prove something, to numb fear with action.

If you want a clean test, try this: write down, in one sentence, why you own each position. Not a story about the future of finance-your reason. If you can't write it, the position is probably not an investment. It may still be a valid choice as entertainment or experimentation, but it should be sized like entertainment, not like a retirement plan.

The uncomfortable truth is that markets don't just price assets. They price people. They price impatience, insecurity, ego, fear, and the human need to feel like you're not falling behind.

The good news is that you don't need to win every battle inside your head. You just need to stop losing the big ones—the ones where you buy because you feel left out, hold because you feel ashamed, and sell because you feel scared.

You can't remove emotion from investing. You can, however, keep emotion from being your portfolio manager. That's the difference between being exposed to risk and being owned by it.

Chapter 2

The Technology Engine: How Tokenization Works

If traditional finance is a manual transmission car driven by intermediaries, tokenization is a self-driving electric vehicle. You don't need to be a mechanic to ride in it, but understanding the engine helps you trust the journey. In this chapter, we strip away the intimidating jargon to look at the three critical components of this engine: the ledger that stores the truth, the smart contracts that automate the rules, and the standards that ensure everything works together. We will explore how we move from a paper deed in a vault to a digital token in a wallet, and why 'programmable money' is the biggest upgrade to finance since the credit card.

2.1 The Digital Ledger: Consensus and Trust

A modern financial system runs on a deceptively simple promise: when the ledger says you own something, you own it. Not "sort of," not "unless there's a mismatch," not "pending someone's spreadsheet correction." In finance, the ledger isn't a record of reality—it *is* reality. Securities ownership, cash balances, collateral positions, beneficial interests in funds: they all reduce to entries in databases that institutions agree to treat as authoritative. The pipes, protocols, and paperwork exist to keep those entries consistent across firms, across borders, and across time.

For most of financial history, the ledger lived in a place. A vault. A registry office. A bank's mainframe. A central securities depository.

Even when everything went digital, the architecture kept the same shape: a hub that everyone trusts, with spokes that connect clients, brokers, custodians, and clearinghouses to the hub. That design has virtues—speed, privacy, governance clarity—but it also has a distinctive vulnerability: whoever controls the file controls the world it describes. In other words, the question "who owns what?" quietly becomes "who is allowed to edit the master record?"

The easiest way to feel the difference is to picture the old model as a shared Excel workbook that only one person can open in edit mode. Everyone else can email requests—"please transfer 100 shares," "please update my address," "please release collateral"—and someone with permission pushes the buttons. Most days, that's fine. On the bad days, it's operational risk in a crisp suit: fat-finger errors, delayed updates, mismatched records between firms, outages, and the awkward truth that if the editor's laptop dies, the business doesn't simply slow down—it can stop.

Distributed ledger technology (DLT) flips the geometry. Instead of one "golden copy" and a swarm of dependent copies, the system is designed so multiple parties maintain the record together. The file isn't "owned" by a single institution in the way a bank owns its core banking database. It's shared infrastructure—like a market utility—except the utility is implemented in software that can run across organizations and, in public networks, across the world.

That sounds like a small technical detail until you translate it into power. In a centralized database, trust flows from an operator. In a distributed ledger, trust flows from a process.

Town hall is the cleanest analogy. Imagine a town where property rights are tracked on a noticeboard in the town square. Every sale, inheritance, or lien gets posted. If someone tries to claim your house, you don't have to chase down a filing clerk in a back room—you point to the board. Now imagine the town doesn't trust a single clerk to maintain it. So dozens of residents take turns copying the board, comparing copies, and agreeing on what the board says *right now*. If one person's copy is smudged or maliciously altered, it doesn't win. It gets outvoted by the consistent majority.

That's the core move: from "trust the keeper of the ledger" to "trust the method by which the ledger is kept."

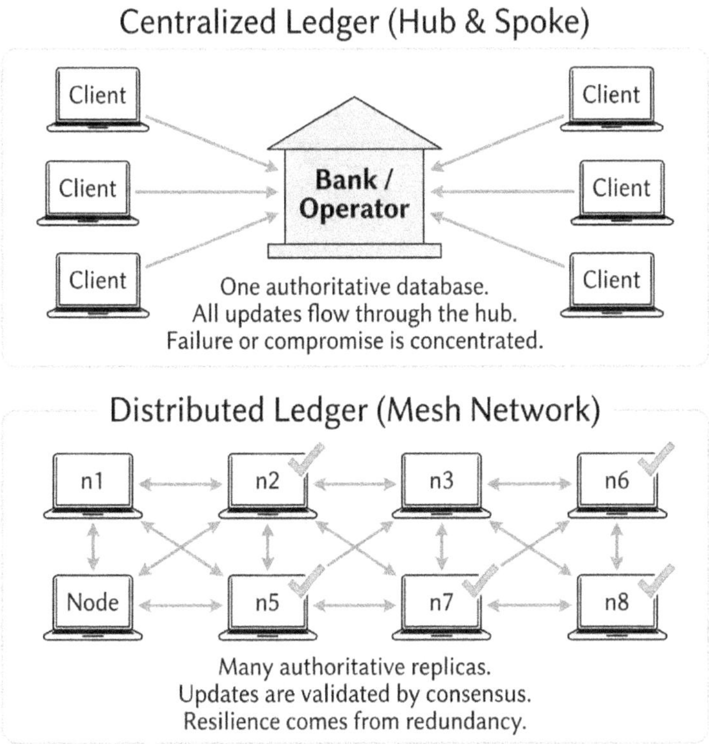

That "agreement" word does a lot of work. In a typical bank database, agreement is easy: the system administrator decides what is true because there is only one system that matters. In a distributed ledger, agreement is the product. It's the output of a mechanism called *consensus*: a set of rules that determines which transactions are valid, what order they occurred in, and what the ledger looks like after applying them.

Here's why order matters. If Alice has $100 and sends $100 to Bob and $100 to Carol, the ledger can't simply store both transactions and call it a day; the order determines which one "wins." Traditional systems solve this with a centralized sequencer: the bank's database assigns timestamps, locks rows, and prevents simultaneous inconsistencies. A distributed ledger has to solve the same problem without a single lock manager. Consensus is the method that makes a globally shared ordering possible.

In finance, this is more than an engineering puzzle. It's the heart of settlement. When markets talk about "T+1" or "real-time settlement," they're talking about when the ledger becomes final—when the entry stops being negotiable and becomes enforceable. Distributed ledgers put that question directly into the design of the network: *how, exactly, does finality happen when no one owns the pen?*

Two broad families of answers dominate.

Public blockchains—open networks where anyone can join—tend to rely on *economic* security. Participants are rewarded for behaving honestly and punished (or simply outcompeted) for trying to cheat. Consensus in this world isn't just software; it's game theory with money at stake. The result is a ledger that is remarkably hard to rewrite, not because rewriting is mathematically impossible, but because rewriting becomes prohibitively expensive.

Permissioned ledgers—private or consortium networks where validators are known entities—lean on *organizational* security. Banks, market utilities, and regulated firms run the nodes. The consensus mechanism can be faster and more deterministic because it doesn't need to defend against an unlimited set of anonymous attackers; it needs to coordinate a defined group with governance rules. You get a different trust profile: less "trust the crowd," more "trust the club, and the controls that bind it."

Both models are attempting the same feat: making the ledger a single source of truth without a single point of control. They simply choose different ways to make misbehavior irrational.

The phrase "single source of truth" deserves a moment, because it's one of those comforting slogans that can hide a mess. In many financial markets today, there isn't a single source of truth—there are multiple sources of "truth," each authoritative *within its own boundary*. Your broker has a record. Your custodian has a record. The central securities depository has a record. The fund administrator has a record. Reconciliation is the daily, expensive ritual of forcing those truths to agree. When mismatches occur, the industry doesn't ask "what does the ledger say?" It asks "which ledger do we trust more, and how quickly can we prove it?"

Distributed ledgers are appealing because they aim to make reconciliation obsolete in the simplest possible way: stop having multiple ledgers. Replace a web of synchronized databases with a shared database. Instead of spending time proving that records match, parties can spend time doing business. It's not that disputes disappear—finance is too creative for that—but the battlefield changes. The arguments move up a level: from "what happened?" to "what did we agree the rules should be?"

So what stops a participant from slipping in a fraudulent transaction? Cryptography is the first line of defense. Not the mystical, black-box kind—cryptography as a function, like a tamper-evident seal on an envelope.

A digital signature is the modern wax seal. When you sign a transaction—"transfer this token," "move this collateral," "update this record"—the signature proves that someone holding a particular private key authorized it. If the message changes by even one character, the signature no longer matches. This is what turns a distributed network of strangers into something that can still enforce basic property rights: only the holder of the key can move the asset. The system doesn't need to know your name; it needs to know that you can produce the correct signature.

Hashing is the seal's companion. A hash function takes any data—an account state, a transaction list, a document reference—and produces a short fingerprint. Change the data, and the fingerprint changes. In ledger design, hashes are used like rivets. They bind blocks of transactions together so that rewriting history becomes visible. You can still attempt to change the past, but you can't do it quietly.

That last word—quietly—is crucial. A distributed ledger is often described as "immutable," but that can mislead. Immutability isn't a magical property bestowed by "blockchain." It's a practical outcome of (1) how the data is chained together and (2) how hard it is for an attacker to convince the network to accept an alternative history. On some networks, "hard" means "need to control a large fraction of the system's validating power." On others, "hard" means "need to corrupt enough known validators to break governance rules." Different ledger designs, different threat models.

The combination of signatures and hashing answers the question "is this transaction authentic and untampered?" Consensus answers a different question: "which authentic transactions count, and in what order?" You need both. A world where transactions are signed but not ordered is a world where double-spending is easy. A world where transactions are ordered but not signed is a world where anyone can spend your money. The ledger's trust engine is built by bolting these pieces together.

Once you see the ledger as a trust engine, an odd fact becomes easier to accept: distributed ledgers don't eliminate trust; they *move* it. Traditional finance asks you to trust institutions—the bank, the exchange, the clearinghouse, the regulator, the courts. DLT asks you to trust the system design—its incentives, its governance, its code, its validators, its operating assumptions, and the ways it can fail. The comforting part is that you can inspect much of that design. The unsettling part is that you may be trusting things you've never had to think about before, like software upgrade procedures or the economics of network participation.

Consider what "resilience" means in each model. In a centralized ledger, resilience is about backups, disaster recovery sites, cyber defense, and operational controls. The failure mode tends to be discontinuous: when the central system is down, everyone waits. In a distributed ledger, resilience is achieved by redundancy: many nodes store and validate the same data. The failure mode tends to be more graceful: if a few nodes go offline, the network continues. That's why the mesh diagram matters. It's not just topology; it's the shape of operational risk.

Transparency is the other headline feature, and it also comes with trade-offs. A public distributed ledger is a noticeboard in the middle of the square: everyone can see the postings. That can be a feature for auditability and market integrity, but it's uncomfortable for finance, where confidentiality is not a luxury—it's oxygen. Even if addresses are pseudonymous, transaction graphs can leak behavior, relationships, and strategy. Firms don't merely want privacy to hide wrongdoing; they want privacy because revealing positions and flows in real time can be economically damaging.

Permissioned ledgers often answer that by restricting who can read the data, not just who can write it. Public ledgers often answer with techniques that minimize what must be revealed while still proving that rules were followed. Either way, the underlying tension is the same: the ledger can be more transparent than traditional systems, but finance often needs selective visibility—auditors can see, counterparties can see what they need, and competitors can't see the playbook.

The most important shift, though, isn't resilience or transparency. It's the change in what counts as *trustworthy bookkeeping*.

In a centralized model, an entry is trustworthy because an institution with legal responsibility maintains it. If something goes wrong, you have a responsible party to call, sue, fine, or regulate. The ledger is a product of authority.

In a distributed model, an entry is trustworthy because it survived a gauntlet: signatures verified, rules enforced, validators agreed, and the network accepted the update. The ledger is a product of procedure.

That difference reshapes the emotional texture of finance. A centralized ledger feels like customer service: someone is accountable, even if it's slow. A distributed ledger feels like physics: the rules execute the same way regardless of who you are, and there may be nobody to bargain with when you don't like the outcome. For markets, that can be a feature—predictability is valuable. For consumers and regulated institutions, it can be a governance challenge: who gets to change the rules when the world changes?

That brings us to the least glamorous but most consequential topic in DLT: governance. Ledgers don't run themselves forever. Software gets upgraded. Bugs get patched. Parameters get tuned. New compliance requirements appear. In centralized systems, governance is straightforward: the operator decides, within the constraints of regulation and contracts. In distributed systems, governance becomes a design choice. Some networks embed governance on the ledger, with formal voting. Others rely on off-ledger coordination—standards bodies, developer communities, consortium agreements, or simply the gravitational pull of what the major participants adopt.

For financial applications, this is where the romance of "trustless" often collides with reality. Regulated finance doesn't just care that a ledger is hard to tamper with. It cares about legal finality, operational accountability, and the ability to respond to fraud, mistakes, or court orders. The ledger may be shared, but someone still has to answer basic questions: Who is allowed to validate? What happens if validators disagree? What happens in an emergency? Who can pause the system, if anyone? How are disputes handled when "the code did what it was told," but the outcome violates policy or law?

Those aren't technical footnotes. They're the bridge between a clever network and something that can safely carry payroll, mortgages, government bonds, and institutional collateral.

And yet, the attraction remains powerful, because today's financial plumbing pays an enormous tax for having too many "truths." Settlement cycles exist partly because it takes time for institutions to exchange messages, reconcile records, manage exceptions, and reduce the risk that someone fails to deliver what they promised. When the ledger itself is shared, messaging and reconciliation can shrink from being major processes to being side effects of a transaction broadcast and accepted by the network.

Another way to say it: the ledger stops being the *result* of back-office processing and becomes the *platform* on which processing happens.

It's also why distributed ledgers are sometimes misunderstood as merely "a database with multiple copies." The multiple copies are only half the story. The other half is the shared rulebook: everyone is not just storing the same data; everyone is running the same verification logic, producing the same result, and rejecting changes that don't fit the agreed constraints. Without that, you just have synchronized spreadsheets—useful, but not transformative.

If all of this sounds abstract, anchor it in the most physical thing in finance: settlement risk. When you buy a bond, two valuable things move in opposite directions—cash and the bond. The safest version of that exchange is delivery-versus-payment: either both happen, or neither happens. Traditional infrastructure achieves this with intermediaries, netting, timing windows, and legal frameworks. A distributed ledger can encode the atomic swap idea directly into the ledger's state transition rules: the update that moves the bond only becomes valid if the update that moves the cash is also valid. No email chain. No manual matching. No "we'll settle tomorrow and hope nothing breaks overnight."

That's the promise. The caution is that technology doesn't abolish risk; it rearranges it. A centralized system concentrates operational risk in the operator. A distributed system spreads operational responsibility across participants and concentrates new forms of risk in the shared protocol: bugs, misconfigured nodes, flawed governance, and consensus failures. When something goes wrong in a distributed system, the question isn't "which bank's database is wrong?" It's "what does the network consider true, and can it be changed?"

This is why cryptography and consensus are framed as a trust engine. They aren't decorative blockchain trivia. They are the machinery that lets a ledger be both shared and authoritative—an unusual combination in finance, where shared infrastructure is common but shared authority is not.

The practical takeaway is simple and surprisingly radical: with a distributed ledger, *ownership becomes a network fact*. Not a statement printed on a monthly report. Not a reconciliation between two institutions. Not a belief that a trusted party maintained a clean database. A network fact: agreed by a defined set of validators (public or permissioned), proven by signatures, anchored by tamper-evident data structures, and replicated across many machines so the record can survive both mistakes and malice.

For tokenized finance, that's the starting gun. Once you can reliably answer "who owns what?" without relying on a single keeper of the book, you can begin to reimagine what the book is capable of doing.

2.2 Programmable Money: Smart Contracts

A ledger that everyone can agree on is already a powerful thing. But it still behaves like a very strict notebook: it records what happened after humans decide what should happen. Smart contracts change the personality of the ledger. The notebook grows arms. It can hold assets in place, check conditions, and move value the moment rules are satisfied—no back office "processing window," no polite emails asking someone to please confirm receipt, no awkward limbo where everyone believes theyŕe settled but nobodyś systems line up.

The phrase *smart contract* is unfortunate marketing. Itś neither particularly smart, nor automatically a legal contract. Itś best understood as a small, persistent financial robot: software that lives on the ledger, can hold and transfer digital assets, and executes the same way for everyone because the network enforces its logic. If a traditional contract is a promise written in legal language and enforced later (sometimes much later) by humans and courts, a smart contract is a promise expressed as code and enforced immediately by the ledgerś rules.

To make it tangible, forget blockchains for a moment and stare at a vending machine. You put in $2. The machine checks that you paid enough. It checks that the item is in stock. It then releases the snack. No cashier evaluates your creditworthiness. No one sends you an invoice and waits for you to pay. The machine is dumb, but it is relentless: same inputs, same outputs, every time. That relentless consistency is exactly what finance pays armies of operations people to approximate.

A vending machine also reveals the real trick: *escrow*. The machine temporarily holds your money and refuses to complete the exchange unless the rules line up. In financial markets, escrow exists everywhere—often disguised as "clearing," "settlement," "margin," "custody," or "pending." Smart contracts are escrow with a spine. They can hold tokens, release them only under specific conditions, and do it without trusting a particular intermediary to behave.

Picture a simple tokenized bond paying interest every quarter. In the conventional world, the bondś "rules" live in a stack of documents and in the operational runbooks of paying agents, custodians, and administrators. In a smart-contract world, some of those rules can become executable: if the date is March 31, June 30, September 30, or December 31, then pay interest to whoever holds the bond tokens at the relevant record time. The ledger doesńt need to send messages across half a dozen systems, hoping they reconcile. It computes, transfers, and records in one motion.

This is the moment where finance people often squint and ask, "So itś just automation?" It *is* automation—but of the most expensive kind: automation of *coordination*. Most operational friction in capital markets isnt́ caused by math being hard. Itś caused by organizations being separate. Each firm has its own systems, incentives, and risk constraints. Smart contracts can shrink the space where coordination errors live by turning multi-party processes into single, shared state transitions.

A useful way to describe a smart contract is: *a program that controls assets*. That single detail is why the stakes are higher than normal software. Plenty of code runs inside banks today—risk models, pricing engines, booking systems. If those systems fail, itś painful but often reversible: you rerun a batch, restore a backup, correct a record. Smart contracts, especially on public networks, can be more like a self-driving car with no steering wheel installed after launch. Once deployed, the contract can be hard or impossible to change, and it may control value directly. A bug can become a balance sheet event.

Smart contracts are also not "apps" in the usual sense. With a typical app, the developer runs the server and can patch it on Tuesday. With a smart contract, the *network* runs the logic, and the network expects the logic to be deterministic: the same transaction must produce the same result for everyone validating it. That requirement shapes what smart contracts can do. They are excellent at enforcing crisp rules about balances, transfers, and permissions. They are terrible at handling ambiguity, nuance, or "reasonable efforts." If your process depends on human judgment, you either keep that judgment outside the contract or you encode a governance mechanism inside it (which is still, ultimately, a rule).

If money is programmable, then the natural language for money becomes "If This, Then That." If payment arrives, then transfer the asset. If a deadline passes, then apply a penalty. If collateral falls below a threshold, then trigger a margin call. If a vote reaches quorum, then update a parameter. In traditional finance, these are implemented through contracts, intermediaries, and operational workflows. In smart-contract finance, they can be implemented as state machines—systems that move from one defined state to the next only when pre-set conditions are met.

This is where the "robot" metaphor earns its keep. A smart contract doesn't get tired at 2 a.m. during month-end. It doesn't misread an email thread. It doesn't decide it's going to take a long weekend. It also doesn't understand context. It cannot be persuaded, shamed, or gently nudged by relationship managers. It will execute the rules you gave it, even if the result is absurd in hindsight. That combination—perfect consistency plus zero common sense—is why smart contracts can be both liberating and dangerous.

The most famous promise is reduced *counterparty risk*. Counterparty risk is the fear that the other side won't do what they promised: they might fail, stall, dispute, or disappear. In many market structures, counterparty risk is managed by inserting an institution in the middle—clearinghouses, custodians, agents—entities whose job is to stand between parties and make performance more reliable. Smart contracts offer a different approach: remove as much discretion as possible from the process. If the rule is "asset moves only when payment arrives," then the contract can enforce delivery-versus-payment automatically. Nobody has to trust the other party's operational competence; they just need to trust the contract logic and the ledger's ability to enforce it.

But "trust the contract logic" hides a problem that experienced finance people spot immediately: who wrote the logic, and how do we know it's right? In traditional markets, a long chain of institutions provides redundancy and control—sometimes clumsy, but often protective. A transfer agent might catch an issue a broker missed; a custodian might block an instruction; an operations team might pick up the phone. Smart contracts can remove that friction, and along with it, some of the soft safety nets.

That's why mature smart-contract design tends to look less like a hacker's weekend project and more like institutional engineering: layered permissions, explicit emergency procedures, conservative defaults, and an obsession with testing. The code isn't simply "the code." It becomes policy. It becomes operational procedure. It becomes a form of market infrastructure.

A subtle but important feature of smart contracts is *composability*. In plain English: contracts can call other contracts, like building blocks snapping together. That allows fast innovation—lending protocols that integrate price feeds, collateral vaults, liquidation mechanisms, and settlement logic in a single automated loop. It's also how complexity sneaks up. In traditional finance, system boundaries are often enforced by corporate boundaries and regulatory perimeters. In smart-contract systems, boundaries are porous. Your "simple" lending contract might depend on an oracle, a token standard, a bridge, and a governance module written by different teams with different incentives. When failures happen, they often look less like a single crack and more like an earthquake along multiple fault lines.

Even when the smart contract itself is perfect, it still lives in a world that contains facts not native to the ledger. Interest rates change. Reference prices move. A borrower's collateral value falls. A bond issuer defaults. These are real-world conditions, and smart contracts don't have eyes. They need a messenger to bring external information into the ledger. That messenger is often an *oracle*—a service that reports off-ledger data to on-ledger code. Oracles are powerful because they connect programmable money to reality. Oracles are also dangerous because they can become a single point of failure or manipulation. If the oracle lies, the robot obeys.

There's another reason smart contracts are often described as "automated financial robots": they don't just move money; they can enforce *governance*. In conventional finance, governance is slow by design. Boards meet. Committees vote. Documentation is prepared. Regulators are consulted. In a smart-contract system, governance can be encoded: token holders vote, a quorum threshold is checked, and if the threshold is met, a parameter changes. That can be efficient. It can also be unnerving, because it compresses the timeline between decision and consequence. In markets, speed is not always the same thing as safety.

All of this raises the obvious question: if smart contracts can do so much, why isnt everything already programmable? Because finance is full of edge cases, and edge cases are where code goes to get embarrassed.

A classic example is reversibility. When a wire goes to the wrong place, banks and payment networks have procedures: recalls, investigations, sometimes legal action. They are imperfect, but they exist. Many smart-contract systems are designed to be final: transactions, once executed, are extremely difficult to unwind. Thats a feature when you want strong settlement finality. Its a liability when users make mistakes or when attackers exploit logic flaws. The more "robotic" the system becomes, the less room there is for human mercy—and finance, despite its reputation, relies on mercy more than it likes to admit.

Another example is upgrades. Traditional software is patched constantly. Smart contracts often cant be patched in place. Some systems use upgradeable contracts—logic that can be replaced under defined permissions—essentially giving the robot a replaceable brain. That solves one problem and introduces another: if someone can upgrade the contract, then users are trusting not only the initial code but the upgrade governance. An upgrade key becomes a kind of master key. Lose it, and you risk catastrophic theft. Use it irresponsibly, and you undermine the whole promise of predictable execution.

This is where smart contracts become interesting for finance not merely as a technology, but as a discipline. They force uncomfortable clarity about controls. In the old world, many controls are social and procedural: segregation of duties, approvals, reconciliations, audits, operational sign-offs. In a smart-contract world, those controls can be translated into explicit permissions and on-ledger checks. Who can move collateral? Under what conditions? Who can pause transfers? What constitutes an emergency? How is an emergency declared? The robot insists on an answer.

Smart contracts also shift where "work" happens. In conventional systems, you can do a lot of work off-ledger and then submit the final result: net positions, end-of-day reconciliations, aggregated transfers. Smart-contract systems tend to make each step explicit and verifiable. That can increase transparency and reduce disputes, but it can also increase operational complexity if not designed carefully. The ledger records not just outcomes, but process. For markets that have been living in a world of partial visibility and private bookkeeping, thatś a cultural change as much as a technical one.

If all this sounds like it belongs in the realm of crypto trading rather than regulated finance, consider a more familiar framing: smart contracts are the software equivalent of a standing instruction at your bank—except enforceable at the asset level, not merely as a bank promise.

A standing instruction says: "On the first of every month, pay rent." A smart contract can say: "On the first of every month, transfer cash tokens from tenant to landlord *only if* the tenantś balance is sufficient; otherwise, apply a late fee and start a countdown that triggers an eviction right encoded elsewhere." Whether that is desirable is a separate debate. The point is that money stops being passive. It becomes conditional.

And conditionality is what turns tokenization from digital paperwork into a different operating system. When an asset is represented as a token, and the token can be controlled by code, you can embed behavior directly into the asset: transfer restrictions, automated compliance checks, corporate actions, collateral rules, and settlement logic. The asset doesnť merely sit in custody waiting for instructions; it participates in its own administration.

The most practical way to think about smart contracts in finance is not as "replacing lawyers" or "eliminating intermediaries," but as changing the division of labor between institutions and infrastructure. Some tasks that used to require a chain of trusted operators can be moved into shared, verifiable execution. That can reduce disputes, shorten settlement cycles, and lower the cost of complexity. At the same time, it concentrates a different kind of responsibility in design and security engineering. In the old world, errors are often caught by humans because humans are everywhere. In a programmable world, errors are caught by *tests*, *audits*, and *incentives*—or not caught at all.

The result is a new kind of financial literacy. Owning programmable assets means understanding not just what you own, but what the code that governs it is allowed to do. When a token sits in your wallet, you might control it—or a contract might control it under certain conditions. When you "deposit" tokens into a protocol, you might be granting permission for liquidation, rehypothecation, fee collection, or governance-driven parameter changes. The asset becomes a relationship with a robot, and relationships are defined by permissions.

That may sound alien compared to familiar brokerage accounts, but the underlying question is old: *what are the rules, and who enforces them?* Smart contracts answer: the rules can be explicit, executable, and enforced by the same consensus that keeps the ledger consistent. When it works well, it feels like finance with fewer moving parts and fewer excuses. When it works poorly, it feels like arguing with an ATM that swallowed your card—except the ATM is holding your collateral, and it has no customer service line.

Programmable money is not magic. It is disciplined automation of financial promises—automation that turns certain kinds of trust into code and certain kinds of operational risk into engineering risk. The promise is speed, precision, and coordination without constant human babysitting. The price is that the robot will do exactly what it's told, and finance will have to become much more careful about what, exactly, it asks for.

2.3 The Tokenization Lifecycle: From Rights to Tokens

Tokenization sounds like alchemy: take a building, wave a blockchain wand, and out pops a tradable token. The real work is less mystical and more like plumbing in an old city. You're not only laying new pipes—you're figuring out which pipes are *actually connected* to the house, who has the legal right to touch them, and what happens when the water pressure changes at 3 a.m. A token can be engineered in an afternoon. A token that faithfully represents real-world rights, survives audits, and behaves predictably under stress is an institutional project.

Start with the uncomfortable truth: most real-world assets don't exist as "things." They exist as *rights*. Your share in a company is a bundle of rights. A bond is a promise with a payment schedule and covenants. A building is a deed, leases, liens, insurance policies, and a long list of ways you can be prevented from doing what you thought "ownership" meant. Tokenization is the act of taking that messy, legal bundle and turning it into something that a digital ledger can recognize, transfer, and settle.

That's why a good tokenization project begins not with code, but with a question that feels almost philosophical: *what, exactly, is the token supposed to be?* Is it a direct claim on the asset? A claim on a vehicle that owns the asset? A contractual claim on cash flows? A participation interest? An IOU issued by an intermediary? Different answers lead to different regulatory classifications, different risk profiles, and different operational responsibilities. The code is downstream.

Asset selection is often described as step one, but it's really step zero: choosing an asset whose rights are sufficiently well-defined that you can map them into a system that prefers crisp rules. Liquid, standardized assets (like certain bonds) are tempting because their life cycles are familiar and their data is structured. Idiosyncratic assets (like a single property with bespoke leases) are tempting because they make for great demos. In practice, the hardest part is rarely the asset's glamour; it's its paperwork. A token can't rescue sloppy documentation. It can only digitize it.

Once the asset is chosen, tokenization becomes a translation exercise between two worlds that don't naturally speak. The legal world speaks in nuance, exceptions, and *reasonable efforts*. The ledger world speaks in state transitions: either the token moved or it didn't. Bridging that gap requires something that many blockchain conversations try to avoid: *legal wrapping*. The wrap is the structure that makes the token correspond to an enforceable claim.

In many real-world tokenizations, the wrap is a special purpose vehicle (SPV)—a legal entity created to hold the asset (or hold the rights to the asset) and issue tokens that represent claims on that entity. The SPV is not a gimmick. It is a way to make the off-ledger world legible. If you can't put a deed "on-chain," you can put the deed into an entity, and put claims on the entity into tokens. Finance has been doing this for decades with securitizations and structured products; tokenization borrows the same legal machinery and gives it a different distribution and settlement layer.

This SPV step is where much of the real risk hides, because it's where the token's promise is either anchored to reality or left floating as a story. Who is the trustee or director? Who has signing authority for the entity? What happens in bankruptcy? Are token holders creditors, equity holders, or something else? Can the asset be encumbered without token holder consent? If the token is marketed as "fractional ownership," is that accurate in law, or merely poetic? The ledger cannot answer these questions. Courts will.

Legal wrapping also forces a decision about *custody*—not the custody of tokens, but the custody of the underlying asset and the documents that control it. A token that represents a bond held at a custodian is different from a token that represents a claim on a bond held by an issuer. The distinction sounds fussy until something goes wrong. In finance, "who holds what" becomes "who gets paid first" remarkably quickly.

After the legal structure comes the interoperability layer: *standards*. Standards are the reason a USB drive works when you plug it into a laptop you've never seen before. Without standards, every device would require a bespoke connector and a custom driver, and the world would be a museum of incompatible cables. Token standards do the same job for digital assets. They define the basic behaviors a token should have-how it moves, how balances are tracked, how permissions work-so wallets, exchanges, and internal systems can support many assets without rewriting their software each time.

In practical terms, a standard is a shared contract between strangers: "If you build your token this way, my wallet can hold it. My trading system can transfer it. My reporting tools can interpret it." This is not glamorous work, but it's where tokenization either becomes an ecosystem or becomes a drawer full of one-off experiments.

Standards also make a quiet philosophical claim: that an asset is not only what it represents, but how it *behaves*. Two instruments might both be "equity-like," but one has transfer restrictions, lock-ups, or compliance checks. One allows delegated transfer permissions; another forbids them. Standardization doesn't remove these differences, but it provides a baseline vocabulary so systems can at least ask the right questions.

Then comes the part people tend to imagine first: *minting*. Minting is the creation of tokens on the ledger-an issuance event that brings a supply into existence and assigns initial ownership. If the legal wrap is the foundation and standards are the blueprint, minting is pouring the concrete. It's also the moment where the project stops being a slide deck and becomes an operational system that can fail in real time.

Minting is never just "create tokens." It is "create tokens in a way that ties them to the real-world cap table or ownership register." If 10,000 tokens represent 100% of an SPV, who attests that 10,000 is the right number? Who prevents an accidental second issuance? Who controls the issuance key? If tokens represent a debt instrument, how are interest and principal obligations represented—through separate tokens, through contract logic, or through off-ledger processes referenced by the token? The ledger can enforce arithmetic; it cannot ensure the arithmetic corresponds to lawful reality.

The minute tokens exist, you confront transfer rules. Real-world assets often have restrictions: securities laws, sanctions lists, KYC/AML requirements, investor eligibility, jurisdictional constraints, lock-up periods, concentration limits. Tokenization doesn't make these constraints disappear; it forces you to decide where they live. You can enforce restrictions off-ledger (through regulated intermediaries who only allow compliant transfers), or on-ledger (by embedding checks in smart contract logic), or through a hybrid. Each choice changes the asset's liquidity and the system's failure modes.

Enforcing rules on-ledger is appealing because it reduces the "I thought it was compliant" problem. If a token literally cannot be transferred to a non-approved wallet, a whole category of mistakes becomes impossible. The downside is brittleness: compliance logic must be correct, upgradable in lawful ways, and aligned with evolving regulation. Off-ledger enforcement is flexible and familiar but brings back operational friction and, in some cases, recreates the very intermediaries tokenization hoped to reduce.

A token's life doesn't end at issuance. Real assets have *servicing*: collecting rent, paying coupons, processing corporate actions, handling redemptions, distributing reports, managing taxes, responding to defaults. Tokenization is often sold as a way to compress messaging, reconciliation, and settlement into a single operation, but servicing can pull you back into the real world. A building doesn't pay rent in tokens by default. Tenants pay in bank money. Someone must collect, account, and distribute. Even in a tokenized world, "someone" might be a servicer, an administrator, or a bank account controlled by the SPV.

Here's where the concept of *oracles* becomes unavoidable. A smart contract can execute perfectly, but it can only execute on information it can see. Oracles are the bridges that bring external data into the ledger: prices, interest rate indices, proof of payment events, proof of delivery, corporate action announcements, even mundane facts like "today is the 30th." Oracles are powerful because they let tokens reflect the world rather than becoming a sealed terrarium.

Oracles also introduce a specific kind of risk: they are an epistemology problem in financial clothing. *How does the ledger know what it knows?* If an oracle reports a price, where did it come from? If it reports that a payment was received, what system observed it? If it reports that a building's insurance is current, who verified it? An oracle can be decentralized (multiple sources, aggregated) or centralized (a trusted provider). Either way, the tokenization project must decide which facts can be treated as authoritative and how disputes are handled when off-ledger reality and on-ledger state diverge.

The lifecycle becomes clearer when you view it as a controlled handoff between domains:

- The off-ledger world defines rights and obligations.
- The legal wrap packages those rights into a structure that can issue claims.
- The token standard defines how those claims behave in digital infrastructure.
- Minting creates the digital representation and assigns ownership.
- Oracles and servicing keep the representation synchronized with reality over time.

The order matters. If you mint first and think later, you'll end up with a token that trades beautifully until someone asks what it *means*, at which point you discover your asset is less a security than a rumor.

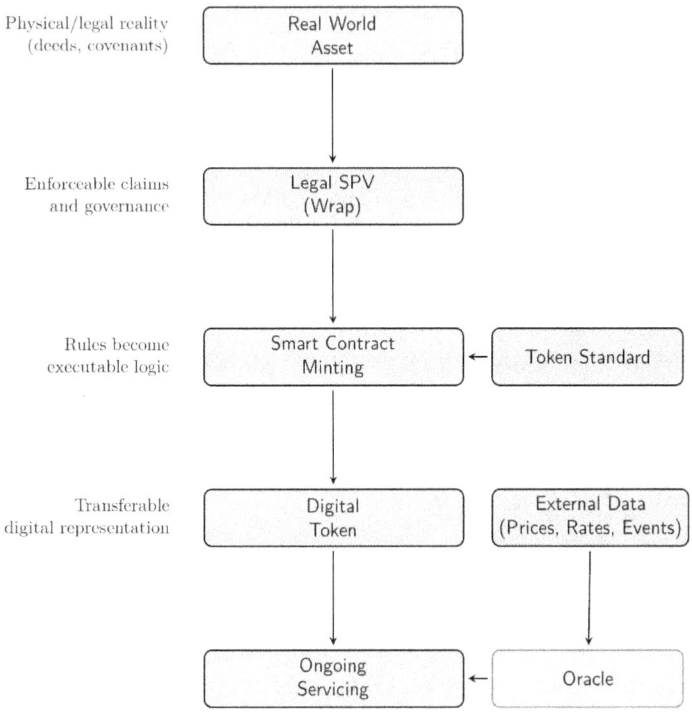

A process map makes tokenization look clean, almost inevitable. Real deployments are messier because institutions rarely flip a switch and declare the new ledger authoritative overnight. Many start with parallel records: the token system records transactions, but the existing system remains the official books-and-records until governance, controls, and regulators are satisfied. That "two ledgers at once" phase can feel like a compromise, but it's often the only responsible way to migrate critical infrastructure. You earn trust by proving, over time, that the new record matches the old one-or, more interestingly, that it produces fewer exceptions.

Parallel operation also reveals something that hype cycles like to ignore: tokenization is not a single event. It is an adoption curve. The first version often resembles the old world wearing new clothes-same intermediaries, same processes, but with a shared ledger used for transparency and reconciliation reduction. Over time, pieces can be re-architected: corporate actions can become more automated, settlement windows can shrink, servicing flows can be more integrated. The destination isn't guaranteed. It depends on incentives, regulation, and whether the operational burden truly drops rather than simply moving to a new team called "protocol operations."

Tokenization is also more than digitization because tokens can carry behavior. A digitized PDF of a bond prospectus is still a document that humans must interpret and enforce. A tokenized bond can have transfer restrictions enforced by code, coupon payments distributed automatically according to holdings, and compliance checks embedded into the transfer function itself. That's a qualitative change: the asset begins to participate in its own administration. It's not just "online"; it is "programmable."

That programmability creates new efficiencies and new risks at the same time. Reducing intermediaries and manual processing can reduce certain operational errors. But composability can create systemic interconnectedness: tokens used as collateral in one place are rehypothecated into another, relying on price feeds and liquidation logic that can cascade under stress. Code risk becomes a form of market risk. An oracle glitch can look like a flash crash. A bug in a widely used contract can behave like a plumbing failure in a major clearing network.

That risk profile is why governance and operational design matter as much as cryptography. A robust tokenization lifecycle includes: clear control over issuance and redemption, well-defined upgrade processes (or clear immutability commitments), transparent audit and monitoring, and explicit procedures for handling off-ledger events like defaults, fraud, court orders, or sanctions updates. The goal is not to create a token that can never fail. The goal is to create a token whose failure modes are legible, bounded, and consistent with the legal rights it represents.

The most revealing question to ask about any tokenized asset is also the simplest: *if something goes wrong, what happens next?* If the underlying asset is damaged, do token holders receive insurance proceeds automatically? If a payment is missed, does the contract freeze transfers? If a regulator demands a transfer restriction, can it be implemented without breaking investor rights? If a private key controlling issuance is compromised, can issuance be halted? If an oracle reports a nonsense price, can the system pause liquidations? Finance lives in its exception handling. Tokenization succeeds or fails there, not in the demo where everything goes right.

When tokenization is done well, the payoff is not merely speed. It's coherence. Messaging, reconciliation, and settlement can collapse into a single, shared update; ownership can be observed and verified rather than inferred and reconciled; corporate actions and servicing can become less manual; and auditability can be improved because the ledger itself becomes a high-integrity record of what happened and when.

When tokenization is done poorly, you get the worst of both worlds: a shiny token on top of a traditional structure, plus new technical dependencies, plus ambiguous legal claims, plus operational teams forced to reconcile not just across institutions but across paradigms. The token becomes a marketing layer rather than a financial primitive.

The lifecycle—from real-world asset to legal wrap to standardized token to minting and ongoing oracle-fed reality checks—is the disciplined path that keeps tokenization grounded. It acknowledges that finance is not just about moving value. It's about defining value, enforcing rights, and surviving the messy, human world where assets live.

2.4 Scale and Privacy: Building for the Real World

The first time most bankers really *feel* a public blockchain, it isn't philosophical-it's physical. You click "send," you wait, you watch confirmations stack up like boarding passes, and you realize you've accidentally time-traveled back to an era when settlement was measured in *patience*. Then you look at the ledger and notice something else: it's not merely slow; it's *public*. Not "auditable by regulators" public. "Anyone with an internet connection can watch the flow" public. Those two traits—throughput constraints and radical transparency—are the reasons tokenization can't simply move the plumbing of global finance onto an early-generation public chain and declare victory.

Speed is the easier objection to understand. Modern payments and trading systems are built to handle the ugly peak moments: market open, payroll day, volatility spikes, holiday catch-up. They're engineered for bursty, industrial-scale throughput. A base-layer blockchain (often called *Layer 1*, or L1) is engineered for something else: making it extremely expensive to change history. That security mindset tends to produce conservative design choices. Every node that validates transactions is asked to do more work. Every block is broadcast widely. Every rule is enforced everywhere. The result is robust agreement—and a hard ceiling on "how many things can happen per second."

That trade-off is not an accident; it's a design philosophy. If your goal is a ledger that strangers can trust without a central operator, you end up paying for redundancy. Thousands of machines repeat the same verification, store the same data, and re-check the same signatures. Traditional finance avoids that by centralizing validation inside a few controlled institutions. Public chains earn trust by *distributing* it. Distribution is reassuring in a geopolitical sense and expensive in a compute-and-bandwidth sense.

There's a second reason L1s feel slow: they are not only transaction processors, they are settlement engines. Settlement is supposed to be boring. It is the armored truck, not the card swipe. An armored truck is secure precisely because it is heavy, cautious, and difficult to hijack. If you try to use it as a high-frequency courier, you either crash the truck or you lower the security standards that made it worth having.

That distinction—execution versus settlement—is how most practical scaling solutions are framed. You keep a secure, widely trusted base layer for final settlement, and you build faster layers on top for day-to-day activity. Those faster layers are called *Layer 2* (L2). If L1 is the courthouse where property rights are finalized, L2 is the bustling market where trades happen all day. You don't want a judge signing each receipt at the farmer's market; you want the judge available when someone disputes the final tally.

Layer 2 scaling is often described as "moving transactions off-chain," which can sound like cheating. But the better mental model is *compression*. Imagine you have a thousand small transfers between the same set of parties. Posting each one individually to a slow, expensive settlement system is wasteful. An L2 can execute those transfers quickly, keep a detailed internal record, and then periodically post a compact proof or summary to L1. L1 becomes the backstop: the place where disputes can be resolved and where final settlement anchors the system.

One popular L2 approach is the *rollup*. The name is literal: roll many transactions up into one package. The details vary, but the core idea is consistent: execute a lot, publish enough information to keep the system honest, and rely on the base layer for security. This is why many scaling roadmaps are described as "rollup-centric"—not because L1 is being abandoned, but because L1 is being treated like a settlement layer should be treated: precious, secure, and not burdened with every coffee purchase.

Even within rollups, the finance-relevant nuance is how trust is enforced. Some designs use validity proofs (often called "zero-knowledge" proofs, though the phrase can mean multiple things): cryptographic evidence that the rollup's state transitions were correct. Others use *optimistic* assumptions: the system assumes transactions are valid unless someone proves fraud within a challenge window. That challenge window is a sober detail with very real user experience consequences. It can introduce delays for certain actions (notably withdrawals back to the base layer) because the system must leave time for disputes. Finance is comfortable with waiting when waiting reduces risk; it's less comfortable when waiting breaks liquidity management.

Scaling also intersects with a more political question: *who gets to be fast?* In centralized markets, speed advantages are often purchased—better connectivity, better colocation, better market access. In public networks, speed can become a competition among validators, sequencers, and transaction senders. The details are technical, but the underlying concern is institutional: if the path to scale creates new gatekeepers (for example, a single sequencer that orders transactions), then part of the original promise—reduced reliance on central operators—returns through the back door. The practical response is not outrage; it's design and governance. Systems can be engineered so sequencing is decentralized, auditable, or replaceable under clear rules. But the point remains: scale choices are also power choices.

Now for the other elephant: privacy.

A public ledger is a strange kind of transparency. It is not like a public company filing a quarterly report—selective, delayed, contextual. It is like having your checking account activity printed on a billboard, except the billboard uses pseudonyms instead of names. Pseudonymity is not the same as privacy. If an address can be linked to an institution (through on- and off-ramp flows, known treasury addresses, counterparty patterns, or simple human error), transaction history becomes a behavioral fingerprint.

Finance runs on confidentiality for reasons that have nothing to do with hiding wrongdoing. Traders protect strategies. Corporations protect payroll data and supplier relationships. Banks protect client balances and positions. Even regulators, who demand transparency, typically want it in the form of controlled access and audit trails—not open broadcasting of sensitive commercial information. A settlement system that leaks positions in real time is not merely "too public"; it can be economically destabilizing.

So how do you get the auditability benefits of shared ledgers without turning every participant into a glass box?

One answer is to avoid fully public ledgers for many institutional use cases and instead use *permissioned* networks. Permissioned systems can restrict who can read and who can write. They can enforce identity at the network layer. They can offer deterministic finality: once a transaction is validated by the known validator set, it is final, not probabilistic. That's deeply appealing in markets where "final" has legal meaning and where operational certainty is a feature, not a philosophical compromise. The trade-off is that you reintroduce reliance on a defined group of operators and the governance that holds them together. You don't eliminate trust; you point it at a different target.

Permissioned design also allows privacy techniques that feel very natural to enterprise engineers. Sensitive payloads can be kept off the shared ledger while a cryptographic fingerprint (a hash) is kept on-ledger. The ledger then proves integrity—"this document existed in this form at this time"—without revealing its contents to everyone. It's the digital equivalent of notarizing an envelope without opening it. For audits and disputes, this can be exactly what institutions want: evidence without exposure.

The other answer is more radical: keep the ledger public but make transactions selectively opaque using cryptography. This is where *zero-knowledge proofs* (ZKPs) enter the conversation, often surrounded by the kind of hype that makes adults roll their eyes.

Strip away the buzz and the concept is wonderfully human: you can prove a statement is true without revealing the underlying secret.

A simple analogy: you want to buy a $900,000 house. The seller wants assurance you can pay. You do not want to hand them your full bank statement, revealing every deposit, expense, and investment. A zero-knowledge proof is the mathematical version of saying, "I can prove I have at least $900,000 available without showing you my total balance or where it came from." The seller gets certainty about the one fact that matters. You keep the rest private.

In finance, that translates into powerful possibilities. You can prove that a transfer obeyed compliance rules—"the sender is KYC'd," "the recipient is not sanctioned," "the position limit won't be breached"—without revealing the parties' identities to the entire network. You can prove that a collateral vault is solvent without revealing every individual position. You can settle transactions with confidentiality while still providing verifiable correctness.

The catch is that privacy is not a feature you sprinkle on top. It changes system design. It affects performance, developer tooling, audit methods, and the way regulators get comfortable. It can also change what "transparency" means. A public blockchain is transparent by default; ZK systems are transparent in the sense that compliance can be proven, but the underlying details are hidden. That is closer to how regulated finance already works: regulators and auditors get visibility, counterparties get limited visibility, and competitors get none.

Which brings us to the paradox at the heart of "blockchain for banking." The most valuable property of public ledgers is open verification: anyone can check the rules were followed. The most valuable property of financial infrastructure is controlled disclosure: only the right people can see the right data at the right time. Bridging that gap is where the real engineering action is happening, and it's why the conversation has moved beyond "public chain versus private chain" into more nuanced hybrids.

For example, you might use a public base layer for settlement finality but keep sensitive trade details off-chain, anchoring hashes for integrity. Or you might use a permissioned ledger for primary issuance and servicing while allowing tokens to interoperate with public ecosystems under constrained gateways. Or you might use an L2 for throughput and privacy, with L1 as a dispute-resolution anchor. These are not ideological compromises. They are attempts to satisfy two masters that both matter: security against manipulation and privacy against commercial harm.

There's also a very practical, very finance-flavored point about privacy: it's not only about hiding numbers; it's about *preventing front-running and information leakage*. In public transaction pools, orders can be observed before they are finalized. If someone can see your large trade about to execute, they can try to jump ahead, pushing prices against you. Markets have wrestled with information asymmetry forever, but public ledgers can create new forms of it if transaction ordering is not carefully designed. The deeper lesson is that market microstructure doesn't disappear when you tokenize assets; it simply migrates into different parts of the stack.

When institutions test tokenized settlement systems—especially in wholesale contexts—the emphasis tends to be refreshingly unromantic: operational integration, safety controls, legal finality, access management, and what happens during outages. Nobody wants a settlement network that is "innovative" in the way an experimental restaurant is innovative. They want it to be dull, predictable, and correct. The interesting part is that tokenization can still deliver meaningful improvements even under those constraints: fewer reconciliations, more atomic settlement, better audit trails, and more programmable workflows—provided the system can scale and preserve confidentiality.

So the "real world" problem of blockchains is not a single problem. It's a bundle of tensions:

- Security loves redundancy; throughput hates it.

- Transparency builds trust; confidentiality enables commerce.
- Deterministic finality is operationally comforting; open participation is politically resilient.
- Programmability removes friction; bugs remove sleep.

What makes the space genuinely exciting is that these tensions are not dead ends. They are design variables. Layering gives a way to separate execution speed from settlement security. Proof systems give a way to separate verification from disclosure. Permissioning gives a way to separate participation from public visibility. Governance gives a way to evolve systems without turning them into private fiefdoms.

A useful rule of thumb for tokenized finance is this: if a proposal requires every trade, every balance, and every instruction to be broadcast to the world in real time, it is probably a toy. If a proposal requires a single operator to be trusted absolutely, it is probably just a database with better branding. The interesting designs occupy the uneasy middle: shared enough to reduce reconciliation and improve integrity, private enough to protect clients, scalable enough to handle real flows, and governed enough to survive both bugs and politics.

Scale and privacy are sometimes framed as "limitations" of public blockchains, as if the technology merely needs to catch up. A better framing is that finance is demanding something unusually hard: a system that is globally verifiable yet selectively confidential, programmable yet safe, fast yet final, open enough to be resilient yet governed enough to be accountable. Traditional infrastructure solves these requirements by layering institutions—banks, custodians, clearinghouses, auditors, regulators—each providing a slice of trust. Tokenized infrastructure attempts to embed more of that trust in the technology itself.

When it works, the payoff isn't just that transactions go faster. The payoff is that markets become less dependent on constant reconciliation, less fragile to operational mismatches, and more capable of atomic, rules-driven settlement. But the only way to get there is to treat throughput and privacy not as afterthoughts, but as first-class engineering requirements—because global finance is not a demo environment, and it does not forgive systems that can't keep secrets or keep up.

Chapter 3

The Digital Asset Taxonomy

If you have ever felt like the crypto world is a chaotic mix of acronyms and jargon, you are not alone. But beneath the noise, almost every digital asset falls into one of four logical buckets. In this chapter, we are going to label the shelves of this digital supermarket so you know exactly what you are buying-whether it is digital cash, a slice of a building, a voting slip, or a one-of-a-kind digital autograph.

3.1 Digital Cash: Payment Tokens and Stablecoins

Money is supposed to be boring. You notice it only when it fails: the card reader goes down, a wire arrives a day late, an account gets frozen at the exact wrong moment. Crypto showed up promising money that works like the internet—always on, globally reachable, indifferent to bank holidays—and then immediately made itself impossible to ignore by swinging 10% in an afternoon. That tension is the starting point for "digital cash": some tokens behave like a risky commodity you can move; others try to behave like dollars you can teleport.

A useful mental split is simple: *payment tokens* are native bearer assets of a blockchain network (Bitcoin being the celebrity example), and *stablecoins* are tokens engineered to hold steady value by anchoring themselves to something off-chain, usually the U.S. dollar. The first category is closer to digital gold bars. The second is closer to digital bank deposits—with a twist: the "bank" is often a non-bank issuer, and the "deposit" is a token that lives on a public ledger.

Bitcoin makes a strange kind of sense once you stop asking it to behave like a checking account. Its price is volatile because its supply rules are intentionally rigid and its demand is emotional, global, and narrative-driven. That's not a bug so much as the point: Bitcoin trades like an asset whose value is a referendum on scarcity, distrust of intermediaries, and the appetite for a bearer instrument you can hold without permission. People call it "digital gold" because gold is also awkward as day-to-day money: you don't pay rent with a bar of metal, but you might sleep better knowing it's there.

This is why "Why can't I buy coffee with Bitcoin?" is both fair and slightly mis-aimed. You *can*. The more revealing question is: why would a merchant price lattes in something that might be worth meaningfully more or less by the end of the week? Volatility turns spending into regret management. If the asset goes up, you feel like you overpaid. If it goes down, the merchant feels like they got underpaid. A medium of exchange wants calm. Bitcoin insists on being dramatic.

Stablecoins exist to remove that drama while keeping the superpower: on-chain transfer. If a payment token is the network's native commodity, a stablecoin is a promise layered on top of the network. The promise is usually plain: "This token is redeemable for one U.S. dollar." That single sentence quietly smuggles in the entire world of balance sheets, custody, audits, and regulation. Stability is not a vibe; it's an engineering claim that depends on who holds what, where, and with what legal obligation.

The cleanest version is a *fiat-backed stablecoin*. "Fiat-backed" means the issuer holds reserves off-chain intended to match the tokens outstanding, typically in cash and short-term, high-quality instruments. When it works well, the stablecoin is boring in the best way: you send it like crypto, you think about it like dollars. If the internet taught consumers to expect instant messaging, stablecoins are an attempt to bring that expectation to money movement.

The mechanism is a loop: dollars go into the issuer's orbit, tokens come out; tokens go back in, dollars come out. The blockchain is the track-and-trace system. The reserves are the substance. The redemption process is the safety valve: it gives sophisticated players a way to pull the token back to its peg when market prices wobble. Without redemption, "$1" becomes a marketing slogan.

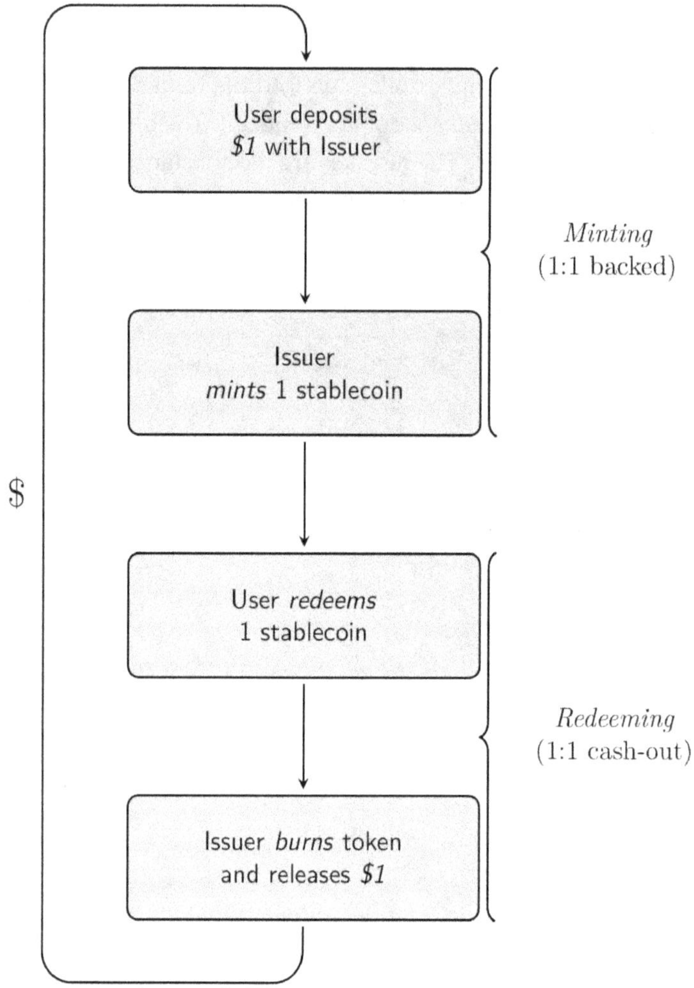

That loop explains why fiat-backed stablecoins often trade very close to $1: if the token slips below $1 on exchanges, a trader who can redeem with the issuer has an incentive to buy cheap tokens, redeem them for dollars, and pocket the difference. If the token rises above $1, the incentive flips: deposit dollars with the issuer, mint new tokens at $1, sell them at a premium, and push the price back down.

In traditional markets, that kind of pressure is called arbitrage. In stablecoins, it's the peg's gravity.

But gravity depends on the mass of the reserves and the physics of redemption. "Redeemable" can mean many things in practice. Some issuers offer redemption only to institutional customers, not retail holders. Some require minimum sizes. Some process redemptions only on business days, and money markets can be closed when crypto trades 24/7. During stress, those frictions matter. A stablecoin can be fully backed and still wobble if the escape hatch is narrow or slow.

Stablecoins are also honest about something banks usually hide behind user experience: money movement is infrastructure, and infrastructure has tolls. On a public blockchain, you pay network fees—often called *gas*—to have your transfer included in the ledger. Gas is not a fee to the stablecoin issuer; it's the cost of using a shared network that doesn't have a customer service line and doesn't do favors. The fee can be pennies or it can be painful, depending on the chain and congestion. Either way, it's the price of independence from a bank's internal ledger.

This is where stablecoins start to feel less like "a crypto product" and more like "a new form factor for dollars." Consider what happens in a normal bank transfer: you instruct your bank, your bank talks to another bank (or a clearing system), balances update, settlement finality arrives later, and both sides rely on a web of rules and recourse. With a stablecoin transfer, the ledger update is public and typically fast. Finality is closer to the surface. The trade-off is that you, not your bank, are responsible for the destination address being correct. There is no "undo." The convenience is real; so is the sharp edge.

Payment tokens and stablecoins also differ in what gives them value. A payment token's value is endogenous: it comes from the market's willingness to hold it and from the network's usefulness and security. Bitcoin does not promise you anything off-chain. It doesn't owe you dollars. It doesn't owe you a yield. It is a bearer asset with a consensus rulebook. That's precisely why it can't "break" the way a badly managed balance sheet breaks; and it's also why the price can float wherever belief and liquidity take it.

A fiat-backed stablecoin's value is exogenous: it borrows credibility from the dollar and the issuer's reserves. The blockchain can be flawless and the stablecoin can still fail if the backing is weak, mismanaged, or legally ambiguous. Stablecoin risk is less about cryptography and more about finance: asset quality, maturity mismatch, concentration of banking partners, operational controls, and the legal structure of redemption claims.

Those finance-native risks show up in very human ways. If reserves are held in safe, short-term instruments, the stablecoin behaves like a well-run narrow bank: dull, liquid, predictable. If reserves stretch for yield or liquidity, you've smuggled a money market fund inside a token wrapper. And if reserves are opaque, you've asked the market to run on trust while advertising "trustless" technology. That contradiction tends to get resolved the hard way.

Even when reserves are strong, stablecoins create a new kind of plumbing risk: the stablecoin may be stable, but the pathway around it may not be. Exchanges can freeze. Bridges between blockchains can get hacked. Wallet software can be compromised. A stablecoin transfer is only as safe as the weakest link in the stack you use to hold and move it. The peg can hold perfectly while a user still loses everything to a phishing link.

The word *stable* is also doing more work than it deserves. Stablecoins are usually designed to track $1, but they are not dollars in the legal sense. They are private liabilities that reference the dollar. That distinction matters in a crisis. Depositors at insured banks benefit from a mature regime of supervision, capital requirements, and (in many jurisdictions) deposit insurance. Stablecoin holders are relying on a different mix of safeguards: reserve management policies, third-party attestations or audits, regulator oversight (where applicable), and the issuer's operational competence. A stablecoin can be well designed and still be a different species than a bank deposit.

So why do stablecoins exist at all if bank deposits already work? Because "work" is contextual. A small business that can't easily access U.S. dollar banking might still access dollar-denominated liquidity via stablecoins. A trading firm can move stablecoins between venues at any hour without waiting for wire cutoffs. A fintech can build programmable money flows—conditional payments, escrow-like behavior, instant settlement of simple obligations—without integrating into every bank's proprietary rails. Stablecoins are less about replacing money and more about unbundling what banks historically packaged together: custody, payments, compliance, credit creation, and user interface.

That unbundling explains why stablecoins often end up as the grease inside crypto markets. Most crypto-to-crypto trading is really crypto-to-stablecoin-to-crypto. Traders want to de-risk without leaving the on-chain environment. A stablecoin becomes the cash leg of the trade, the unit of account for quoting prices, and the waiting room between risk-on positions. It's the closest thing the crypto ecosystem has to "cash management," even if the cash management tool is a token.

Meanwhile, payment tokens like Bitcoin sit in a different psychological and portfolio bucket. People may use them for payments, but many hold them as an asset with asymmetric upside, censorship resistance, or diversification appeal. That investor posture feeds volatility, which then feeds the investor posture. The result is a loop of its own: not a mint-and-redeem loop, but a narrative-and-liquidity loop.

For everyday users, the most important practical distinction between these two kinds of "digital cash" is what happens when you press "send." When you send a stablecoin, you are mostly taking *payment rail risk*: network fees, confirmation times, the possibility of sending to the wrong address, and the operational risk of the wallet or exchange you use. The value you receive (ideally) stays near $1. When you send a payment token, you are taking all of that rail risk *plus* price risk. The transfer might be flawless and still be economically different at arrival.

That difference changes behavior. If you're settling an invoice, paying a contractor, posting collateral, or moving money between your own accounts, stability is the feature. If you're allocating to an asset because you believe its long-term purchasing power will rise, volatility is the admission ticket. People often talk past each other because they're trying to use one tool for the other tool's job.

There's another subtle point that matters in finance: stablecoins can make dollars *portable*, but they do not automatically make dollars *productive*. A stablecoin balance typically does not, by itself, earn the interest rate you might earn in a bank account or a money market fund. The issuer may earn interest on the reserves; whether any of that is passed through to holders depends on the product design and regulatory posture. This is why "stablecoin yield" offerings quickly become complicated: once you promise yield, you've walked into the world of securities, lending, and credit risk. The simple promise "$1 in, $1 out" is powerful precisely because it is simple.

None of this is an argument that stablecoins are risk-free or that payment tokens are impractical. It's an argument that digital cash comes in two very different flavors, and confusing them produces bad decisions. Treating Bitcoin like a checking account leads to stress and sticker shock. Treating a stablecoin like a sovereign currency leads to complacency about issuer and operational risk.

A good rule of thumb is to ask, "Where does the $1 live?" For a fiat-backed stablecoin, the answer is: in off-chain reserves held somewhere in the traditional financial system, under a legal structure that defines who has a claim and how redemption works. For Bitcoin, the answer is: it doesn't. Bitcoin is not a claim on dollars. It's a scarce digital bearer asset whose value floats. One is a promise attached to a ledger. The other is an asset native to a ledger.

Once you see that, the rest of the ecosystem becomes easier to read. Stablecoins are the translator between bank money and blockchain money. Payment tokens are the native fuel and, sometimes, the native savings asset of a networked world. Both can be used to pay. Only one is designed to keep you from thinking about price at all—and when money is doing its job, that's the highest compliment you can give it.

3.2 Bridging the Gap: Security Tokens and RWAs

A blockchain can move value with the indifference of a conveyor belt: it doesn't care whether the "thing" being moved is a meme coin, a dollar-pegged token, or a claim on a Manhattan building. Finance, unfortunately (and necessarily), cares a lot. It cares about *what* the claim is, *who* stands behind it, *what happens if something goes wrong*, and *which court* gets to decide. Security tokens and tokenized real-world assets (RWAs) are where the blockchain's elegant conveyor belt collides with the messy but essential machinery of law.

The pitch is intoxicating: imagine buying $50 worth of a skyscraper, swapping Treasury exposure as easily as sending an email, or using a token as collateral that settles instantly instead of "T+1" (or worse, "it'll be final tomorrow, probably"). Tokenization promises to do for assets what shipping containers did for freight: standardize the package so the world can move it cheaply, reliably, and at scale. The container didn't eliminate ports, customs, insurance, or laws of the sea. It made them easier to interface with. RWAs aim for the same trick.

Start with the least glamorous but most decisive fact: a blockchain token cannot *reach out* and grab a building, a bond certificate, or a bank account. Code can't repossess a condo. It can't compel a tenant to pay rent. It can't force an issuer to honor a coupon payment. The magic ingredient is not the token; it's the *wrapper*—a legal structure that makes the token economically meaningful in the non-digital world. Without a wrapper, "tokenized real estate" is often just a digital souvenir that *refers to* a property, not a claim *on* it.

This is why many serious tokenization projects look less like a DeFi experiment and more like traditional structured finance wearing a new outfit. An asset is placed into a legal vehicle (often a special purpose vehicle, or SPV). That vehicle issues securities or participations. The blockchain token then represents the investor's interest in that vehicle, under a defined set of rights: ownership, dividends or interest, redemption terms, voting, transfer restrictions, and disclosures. The token becomes a new *representation layer* for familiar instruments.

A helpful way to think about RWAs is as a three-layer sandwich. The bottom is stubborn reality: a building, a loan portfolio, gold in a vault, a Treasury bill, a private credit facility. The top is the part the internet can hold: a digital token, recorded on a ledger, easily transferrable, programmable, and (in good designs) auditable. The middle is where the real bridge sits: the wrapper that binds the top to the bottom in a way a judge will recognize.

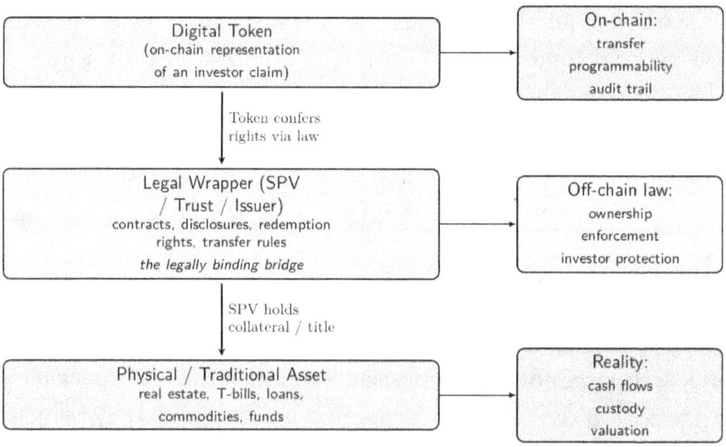

Once that middle layer clicks, a lot of confusion disappears. People sometimes ask whether a token is "real" because it lives on-chain. The more precise question is whether the token is *enforceable*. Enforceability is a legal property, not a cryptographic one. If the SPV is properly formed, if it actually owns the asset, if investor rights are clearly documented, and if transfers are handled in compliance with the rules, then the token can represent something quite serious. If any of those links are weak, the token might still trade, but it's trading on hope and marketing rather than hard claims.

This is why the term *security token* matters. A security token is, economically, what it sounds like: a token that is a security. That's not a slogan; it's a regulatory classification with teeth. Securities law is designed around information asymmetry and human nature—the fact that sellers know more than buyers, and that buyers will happily confuse "exciting" with "safe" when a chart is going up. If a token gives you profit rights, interest, dividends, or a claim on an enterprise managed by someone else, regulators will often treat it like any other security, regardless of whether it is issued on paper, in a brokerage account, or as a smart contract.

The practical implication is almost disappointing: tokenization doesn't repeal compliance. It relocates it. KYC/AML checks still exist. Offering documents still exist. Transfer restrictions still exist. Custody rules still exist. In many tokenized offerings, the investor experience is "traditional finance with better plumbing." The buyer might interact with a wallet, but the wallet address is whitelisted, the transfer is restricted, and the issuer retains controls that look suspiciously like the ones finance has always used.

That disappointment is also the point. Finance is not a video game economy where you can patch the rules after launch. It is a credibility business. Tokenization that tries to dodge the credibility mechanisms tends to discover why those mechanisms were built in the first place—usually during the first real downturn.

So what does tokenization actually improve? A lot, when it's done honestly.

One improvement is *settlement design*. Traditional markets often separate trading from settlement. The trade happens now; the actual exchange of asset for cash finalizes later through a chain of intermediaries. Tokenized instruments can be designed so that delivery and payment occur in a tighter loop, potentially reducing counterparty exposure and operational complexity. That doesn't mean settlement is magically "risk-free"—operational failures, smart contract bugs, and custody issues can still bite—but it can shift risk from opaque plumbing to visible rules.

Another improvement is *fractionalization*, which is a fancy word for "smaller pieces." A $10 million asset can be split into many units without inventing a new share registry process from scratch. This is where the Manhattan-skyscraper thought experiment comes from. But fractionalization is not automatically democratization. If the asset is a security, then securities rules still govern who can buy, how it is marketed, and what disclosures are required. Many tokenized RWAs are offered only to institutional or qualified investors for exactly that reason. The tokens may be divisible; the investor base may not be.

A third improvement is *programmability*. Tokens can carry rules: transfer restrictions, automated coupon payments, corporate actions, or compliance checks embedded into the instrument's lifecycle. Programmability is alluring because it suggests fewer manual reconciliations and fewer "please email operations" moments. It can also be dangerous. When rules are automated, they execute during stress at machine speed. If the rules were designed for sunny days, the system can amplify rather than absorb shocks.

Behind those improvements is the unglamorous reality that tokenization is often a project of *market infrastructure*. It's a way to re-architect registries, settlement systems, and collateral flows. In other words, it targets the parts of finance that are most expensive precisely because they are invisible: recordkeeping, reconciliation, and the endless work of making sure everyone's spreadsheet agrees.

Now, what counts as an RWA? The label is used loosely, but the core idea is simple: something whose value and cash flows originate off-chain is represented on-chain. That can mean tokenized real estate interests, yes. It can also mean tokenized funds holding government bonds, tokenized invoices, tokenized commodity claims, tokenized private credit, or tokenized deposits in certain designs. The "real" part is not a moral judgment about seriousness; it's a reminder that the asset has a legal and economic life outside the blockchain.

That outside life brings outside risks.

Valuation becomes a two-world problem. On-chain tokens can trade continuously, but the underlying building might be appraised quarterly, and the private credit portfolio might be priced based on models and manager marks. The token's price can move faster than the truth underneath it. This is not new—closed-end funds and REITs have lived with similar dynamics—but tokenization can make the mismatch more visible and more tempting to arbitrage.

Liquidity is another trap. Tokenization can make an asset easier to trade mechanically, but it can't conjure natural buyers. A thinly traded token is still thinly traded, just faster. In a crisis, "24/7 liquidity" can mean "24/7 price discovery," which is a polite way of saying the market can reprice brutally at any hour.

Operational risk changes shape. Instead of a single transfer agent or a central securities depository, you might have smart contracts, wallet infrastructure, key management, and possibly cross-chain bridges. The weakest link matters. A tokenized Treasury exposure can be economically conservative and operationally adventurous, and those adjectives do not cancel each other out.

Interconnectedness grows quietly. Tokenization makes it easier to use the same collateral across platforms, to rehypothecate (re-pledge) assets, or to layer leverage through automated lending protocols.

That can make markets more efficient on calm days and more correlated on chaotic ones. When everything settles faster, everything can also unwind faster.

This is why the most credible tokenization efforts tend to start with assets that already have well-understood legal characteristics and deep traditional markets: government bonds, money market fund-like exposures, and other high-quality instruments. The logic is almost conservative: if you're going to introduce a new rail, start with cargo that doesn't explode. Tokenized government-bond exposure, for example, is not interesting because government bonds are new. It's interesting because the bond market's collateral utility is enormous, and anything that makes collateral more mobile can reshape how liquidity moves.

It also explains why a growing number of institutional pilots have used tokenized fund interests rather than "direct" tokenized bonds. A fund is already a wrapper with established governance, reporting, and custody arrangements. Putting fund shares on-chain is often simpler than rewriting the issuance, settlement, and corporate-action machinery of the bond itself. In practice, tokenization frequently enters through the side door: not by replacing an asset, but by tokenizing a familiar vehicle that holds the asset.

This brings us to the most important mindset shift: tokenization is not an argument about whether blockchains are "better databases." It's an argument about *where trust sits*. Traditional finance centralizes trust in regulated intermediaries and legal enforcement. Public blockchains distribute trust across consensus and transparency, then reintroduce centralized trust wherever the real world demands it: custody of reserves, management of collateral, identity checks, and governance. Security tokens and RWAs are the honest middle ground. They admit that code is not a court, but they still try to harvest the efficiencies of a shared ledger.

The best way to tell whether an RWA project is mature is to ignore the token and read the fine print around the wrapper. Who holds title to the asset? Where is it custodied? What happens in bankruptcy? Are token holders senior, junior, or simply "customers" with a contractual claim? Can redemptions be suspended? Are transfers restricted, and if so, how is that enforced? If the answers are vague, the token is likely doing theater rather than finance.

When those answers are crisp, tokenization becomes less of a speculative story and more of a practical tool. A token can function as a programmable share certificate. A blockchain can function as a near-real-time registry. Smart contracts can function as automated operations. And the wrapper can function as the bridge that keeps the whole structure from collapsing into a purely on-chain fiction.

If payment tokens taught the world that value can move without a bank, RWAs teach a more difficult lesson: value can be *packaged* to move without rewriting the world's legal systems. That's not as romantic as "disintermediation." It's more powerful. It's how blockchains start swallowing traditional finance: not by eating the assets themselves, but by eating the workflows that keep the assets in motion.

3.3 Access and Governance: Utility Tokens

A utility token is easiest to understand when it's doing something unglamorous: paying a toll. Not a toll on a bridge made of steel, but a toll on a bridge made of computation. You want the network to run your transaction, store your data, or execute your smart contract; you pay the network's fee token to get in line. That fee isn't a dividend, and it isn't a claim on a reserve account. It's closer to the token you buy at a subway station: you're not investing in the transit authority, you're buying passage.

That framing immediately separates utility tokens from the two categories people most often confuse them with. A payment token can be money-like in the sense that you can hand it to someone else and they'll accept it as value. A stablecoin tries to behave like a dollar on rails. A utility token is different: it is purpose-built for a specific system. Its value comes from usefulness inside that system, not from a promise to redeem it for cash and not necessarily from broad "money-ness."

Unfortunately, the word *utility* has been used like a magic spell. During bull markets, projects have treated "utility token" as a get-out-of-regulation card: *It's not a security, it's a utility.* Markets, being markets, didn't care. They bought because they expected the token to appreciate. Regulators, being regulators, cared very much. They asked a blunt question: are people buying this because they want to use the network, or because they expect profits based on someone else building the network for them?

That question is not philosophical; it's practical. If the token's primary economic reality is "buy now, team builds later, token goes up," it begins to look less like a subway token and more like a fundraising instrument. Securities law is built to deal with fundraising instruments. Calling it "utility" doesn't change the behavior it incentivizes.

So it helps to split utility tokens into two everyday metaphors, neither of which is perfect but both of which are useful: *keys* and *votes*. Keys unlock access to a service or network. Votes steer how that network evolves. Many tokens try to do both at once, which is where things get spicy.

Keys are the simpler story. Some tokens function as prepaid access to a product: cloud storage, compute time, bandwidth, content, or a specialized platform. In that world, a token is basically a metered license. The token might be burned (destroyed) when you use it, like a ticket that gets torn, or it might be held as a membership pass, like a key card that grants ongoing entry. Either way, the token is meaningful because the service exists and people want it.

When the service doesn't exist yet, the "key" story turns into a time machine. You're not buying access; you're buying a placeholder for access. That placeholder might eventually become a real key, but in the meantime it trades like a speculative asset. In that phase, the token's price is not anchored by how much compute or content it can reliably buy today. It's anchored by expectations, and expectations are where financial regulators start reading the fine print.

A network-fee token like ETH shows the key idea in its most mechanical form. On Ethereum, *gas* is the fee paid to include transactions and run smart contract computation. ETH is the token used to pay that gas. This is utility with a hard edge: if you don't pay, the network doesn't execute your instructions. The token is required for the system to function, and demand for it is linked—imperfectly but meaningfully—to the network's usage.

That "imperfectly" matters. A token can have genuine utility and still be volatile. Usage demand can be lumpy, fee markets can spike, and macro sentiment can swamp fundamentals. A subway token would also be volatile if you could trade it globally and if millions of people argued daily about whether subways represent the future of civilization. Utility does not equal stability; it equals *purpose*.

The fee-token design also reveals a deeper economic truth about public blockchains: they are not free to run, and they can't pretend otherwise. Someone provides computing power, storage, and security. The network needs a way to ration scarce block space and to compensate the parties maintaining the system. The fee token is how the network prices itself. In traditional finance, the rationing happens through account access, credit lines, and operational policies. On-chain, the rationing happens through fees.

This brings up a subtle but important difference between utility tokens and normal software pricing. With a typical SaaS product, the company charges you in dollars, pays its cloud bill in dollars, and perhaps offers volume discounts if it wants to. With a tokenized network, the pricing instrument is itself traded in open markets. That means the "price" of using the network has two moving parts: the fee level (gas units) and the token's exchange rate versus the dollar. Users experience that as a kind of double volatility: not only can congestion make fees jump, the token's price can change at the same time. A token that makes sense for network economics can still make users long for a predictable invoice.

Some designs try to soften this by making fees payable in stablecoins or by introducing mechanisms that smooth fee volatility. Those designs are not just technical tweaks; they are choices about who bears risk. In a pure fee-token model, users bear token-price risk. In a stablecoin-fee model, someone else has to manage how the network's economics still work—often by converting stablecoins into the asset that ultimately compensates validators or by maintaining separate incentive structures. The details matter because "fees" are not an afterthought; they are the business model of the infrastructure.

Now consider utility as *membership*. A token might grant access to a community, a content library, an API, a game economy, a trading platform, or a piece of digital real estate. This can feel silly until you remember how many businesses are already membership businesses: gyms, streaming services, airport lounges, professional associations. A token is simply a transferable membership card, sometimes with perks encoded in smart contracts.

Transferability is where tokens become financial objects. A normal membership is typically non-transferable, and if it is transferable it's usually clunky to transfer. A token membership can be sold instantly. That creates a market price for "access," which can be healthy (a transparent signal of demand) or toxic (a barrier that prices out the very users the product needs). A service can accidentally become a speculative asset class. If the token price rises faster than the product's value, the community starts to talk like investors instead of users. That is a cultural shift, and it can kill products.

The other side of utility tokens is *governance*: the idea that holding a token gives you a vote. Governance tokens are often pitched as "ownership" of a decentralized organization, but the resemblance to ownership is sometimes cosmetic. A shareholder vote in a corporation is embedded in a full legal system: fiduciary duties, disclosure requirements, enforceable rights to cash flows, and remedies when management misbehaves. A governance vote on-chain is a rule in software: it can change parameters, direct a treasury, upgrade contracts, or modify incentives. The vote can be powerful while still offering no legal claim to assets, no guaranteed profit rights, and no traditional investor protections.

That sounds like a downgrade, but it's also a different kind of power. A governance token can directly control the behavior of a protocol that moves billions of dollars of value, sets interest rates in a lending market, or allocates rewards to different participants. That is not "just community." That is policy. The token is a voting slip for a miniature economy.

Of course, voting systems inherit the oldest problem in democracy: turnout and capture. In many token-governed systems, most holders don't vote. Some can't be bothered. Some don't understand the proposals. Some hold tokens on exchanges that don't support governance. The result is that a small number of highly motivated, highly resourced participants can steer decisions. On paper, it's one-token-one-vote. In practice, it can be one-whale-one-policy.

The mechanics of governance can make this worse. If votes are weighted by token balance, the system is plutocratic by design: influence scales with capital. Some communities accept this as realism: those with the most at stake should have the most say. Others try to blunt it with delegation (vote-by-proxy), quadratic voting experiments, or reputation systems. Each fix introduces its own complexity and attack surface. Governance is not solved by a whitepaper; it is lived over years, in the unromantic grind of proposals, arguments, and compromises.

Then there's the most awkward feature of governance tokens: the token itself trades. In a corporate setting, you don't buy votes for next Tuesday's board decision unless you're a hostile acquirer. In token governance, you can. That makes governance vulnerable to time-limited accumulation: buy tokens, push through a proposal that benefits you, sell tokens. Some protocols attempt defenses: voting delays, timelocks before changes take effect, quorum thresholds, or "rage quit" mechanics that allow dissenters to exit with their share of treasury assets (where feasible). These are financial-engineering responses to a governance problem, which tells you something about how quickly software governance becomes market governance.

There's also a psychological trap: governance creates the illusion that every choice can be made by vote, but most complex systems need clear responsibility. When everyone is "in charge," no one is accountable. Many successful protocols end up with a hybrid reality: a core team or foundation does much of the work, while token holders ratify big decisions or set broad parameters. That's not hypocrisy; it's organizational gravity. The question is whether the system admits that reality and designs for it, or pretends it doesn't exist and leaves power in informal channels.

So where do utility tokens fit in a digital-asset taxonomy that aims to be practical, not ideological? Think in terms of *what the token lets you do today* and *what people are told it will do tomorrow*. A token that is usable at issuance to pay fees, access a live service, or participate in a functioning governance process has a credible utility story. A token sold widely before those utilities exist is asking buyers to finance construction. Finance can be legitimate, but it is regulated for a reason.

That lens also helps decode hybrid tokens, which are common. A token can be required to pay fees and also marketed as scarce and "number go up." A token can grant access and also entitle holders to revenue sharing. A token can govern a protocol and also serve as collateral in lending markets. The more financial the economic promise becomes—yield, dividends, buybacks, revenue rights—the more the token behaves like a security even if it still has some utility features. Utility doesn't disappear; it just becomes secondary to the investment character.

This is why the phrase "not a security because it has utility" is often a category error. Plenty of traditional securities have utility. Stock in a company can grant access to shareholder meetings and influence over management. Bonds can be used as collateral in repo markets. None of that makes them not securities. What matters is the nature of the claim and the expectations created in the offering and ongoing promotion.

Utility tokens also force an uncomfortable question: what does "ownership" mean in a digital economy? People are used to software being rented, not owned. A token can feel like ownership because it's held in your wallet, transferable without permission, and not tied to an email address. That is a real shift in user agency. But agency is not the same as a legal right to profits, and it is not the same as a guarantee of future value. Many holders learn this distinction only after they've already paid for the lesson.

At their best, utility tokens are a clever way to align a network's users with the network's economics. They can turn usage into participation, participation into stewardship, and stewardship into resilience. At their worst, they are a fundraising tool wearing a gamer skin, selling the romance of "community ownership" while concentrating control and risk in predictable places.

A final, grounding thought: the most durable utility is the kind you can measure without a price chart. If a token reliably buys computation, storage, bandwidth, or access people actually want, it has a reason to exist that survives mood swings. If a token's main "utility" is that it might appreciate because other people might want it later, that's not utility. That's the oldest product in markets: a bet on other people's beliefs, packaged as technology.

3.4 Unique Digital Property: NFTs

If a stablecoin is a digital dollar bill and a security token is a digitized claim wrapped in legal scaffolding, an NFT is something stranger:

a receipt that can't be photocopied. That sounds modest, almost boring, until you remember how much of modern life is made of receipts. Tickets, credentials, warranties, titles, certificates of authenticity, membership cards, "I was here" proofs, "I own this" proofs—the world runs on little documents whose whole job is to settle an argument later.

NFTs (non-fungible tokens) are built to settle a specific argument: *which exact item is yours*. "Non-fungible" simply means not interchangeable. One dollar is interchangeable with any other dollar; one share of a widely traded stock is interchangeable with any other share of the same class; one Bitcoin is interchangeable with any other Bitcoin. An NFT is the opposite. It's closer to a concert seat assignment, a deed to a particular apartment, or the serial number on a luxury watch. The uniqueness is not a marketing flourish; it's the core feature.

The internet has always been embarrassingly good at copying. Copying is not theft in a technical sense; it is the basic function of networks. When you "send" a photo, you don't actually move it like a physical object. You duplicate it. That's why digital scarcity has historically been a social agreement enforced by platforms: your airline ticket is "unique" because the airline's database says it is, and because the bouncer's scanner checks that database. NFTs propose a different enforcement mechanism: uniqueness anchored to a public ledger, where the identifier is difficult to forge and easy to verify.

This is where the million-dollar JPEG era did genuine damage to understanding. People heard "NFT" and thought it meant "image." Often it doesn't. An NFT is a token record on a blockchain, typically following a standard format (commonly ERC-721 for one-of-one items and ERC-1155 for semi-fungible or multi-quantity items). The token can *point* to an image, but the token is not the image any more than a warehouse receipt is the warehouse. That distinction becomes painfully important when links break, metadata changes, or buyers discover they purchased a token that references a file stored somewhere else entirely.

At the heart of an NFT is an ID and some associated metadata—a description that tells software what the token represents and where related content lives. The ledger stores the ownership trail: which address minted it, which addresses held it, and who holds it now. That ownership trail is the real product. When people say NFTs are "digital collectibles," what they often mean is "collectibles with a public provenance record." Provenance is an old obsession in art and luxury goods: who owned it, when, and can you prove it? NFTs make provenance native to the asset.

Provenance is also the bridge from "cartoon apes" to the actually useful stuff. A collectible image is the shallow end of the pool because it's easy to understand and easy to hype. The deeper end is identity and credentials. A concert ticket is a credential: it says you are entitled to entry. A diploma is a credential: it says you completed a program. A professional license, a warranty registration, a membership badge, a supply-chain certificate, a gaming item with scarcity guarantees—these are all claims about *a specific right* attached to *a specific identifier*. NFTs are a general-purpose container for that pattern.

But an NFT can only prove what the system agrees it proves. The blockchain can say "address X owns token Y." It cannot, by itself, force the world to treat token Y as a valid ticket, a valid diploma, or a valid warranty. Somebody has to recognize it. That recognition can be provided by a venue, a university, a brand, a government agency, or a consortium of them. In other words: NFTs feel like pure on-chain magic, but most high-value uses are actually hybrid systems where the NFT is the public handle and the off-chain institution supplies the legitimacy.

This makes NFTs feel less like "art" and more like "infrastructure for claims." It's a subtle shift, but it's the difference between a novelty and a tool. When a ticket is an NFT, resale can be governed by rules embedded in a smart contract: caps on prices, restrictions on transfers, royalty-like fees that route to the issuer, or the ability to revoke a stolen ticket and reissue a fresh one. That last feature sounds like heresy to people who want pure bearer assets, but it is exactly what consumer protection looks like in ticketing. A totally irreversible, totally anonymous ticket is great for scalpers and thieves. It's not obviously great for fans.

The same tension shows up in identity. People hear "NFT as identity" and imagine a permanent tattoo of personal data on a public ledger. That is not what sensible systems do. A well-designed credential scheme can keep private data off-chain and use the NFT as a pointer or proof-of-membership token, while the actual sensitive details live in secure systems or are revealed selectively. Think of the NFT as the badge number, not the full personnel file.

Still, "badge number" can be enough to create new kinds of risk. If an NFT is used as a credential and it sits in the same wallet you use for financial activity, you've combined identity and money in a way that can be convenient and dangerous. Convenience is obvious: single wallet, multiple functions, easy proofs. Danger is quieter: you can accidentally dox yourself by using the same address for public credentials and private wealth. The more NFTs are used for identity-like purposes, the more users will need wallet hygiene—separate addresses, separate devices, separate mental compartments—the digital equivalent of not carrying your passport and your entire net worth in the same pocket.

Another point of confusion is rights. Buying an NFT typically does not automatically buy the intellectual property (IP) rights to the underlying artwork or brand. Owning an NFT that references an image is not the same thing as owning the copyright to that image. It is closer to owning a signed print or a numbered collectible: you own *that item*, not the right to reproduce it commercially. Some projects grant broader licenses, some grant none, and some are intentionally ambiguous. If you're a buyer, "What rights do I get?" is a better question than "Is it on-chain?"

Ambiguity around rights is not just a consumer issue; it's a financial one. A token that confers a legally enforceable bundle of rights can be valued with familiar tools: discounted cash flows, comparable transactions, credit analysis. A token that confers vibes, social status, and optionality is valued like a collectible: sentiment-heavy, reflexive, and prone to booms and busts. Both can trade. Only one is predictable.

And that brings us to regulation, because NFTs sit on a fault line. Technically, an NFT is just a token standard. Economically, it can be almost anything. An NFT can be a ticket. It can be a membership pass. It can be a deed-like record. It can also be sold in a way that looks a lot like an investment contract: marketed with promises of profit, supported by a team whose ongoing efforts are expected to raise resale prices, and distributed to buyers who are not primarily buying for use but for appreciation.

This is where the "it's just a JPEG" defense collapses. Securities analysis tends to focus on substance: what is being offered, what are buyers led to expect, and who is doing the work that might generate profits. If an NFT is pitched like a speculative investment—complete with roadmaps, "floor price" obsession, and a clear expectation that the issuer's future actions will drive value—it starts behaving like a capital raise, even if each token is technically unique. Uniqueness does not immunize an offering from looking like a security.

The market has already supplied case studies of how quickly NFTs can drift from "digital receipt" into "financial product." When buyers are promised exclusive access *and* the issuer emphasizes the secondary market *and* the marketing centers on potential appreciation, the NFT becomes a kind of hybrid: part membership, part speculative instrument. This isn't automatically illegal or illegitimate, but it does change the regulatory perimeter and the risk profile. The buyer is no longer just purchasing an object; they are purchasing exposure to a project team's execution risk and to the social dynamics of a resale market.

Even outside of explicit securities questions, NFTs expose a uniquely modern kind of operational fragility: link rot. Many NFTs refer to metadata hosted off-chain. If that hosting disappears, changes, or is compromised, the token can remain on-chain while the thing it references dissolves into a 404 error or mutates into something else. Some projects mitigate this by storing content on decentralized storage networks or by embedding content more directly, but trade-offs remain: permanence costs money, and many systems optimize for cheap minting rather than durable archives. A "permanent" token pointing to an impermanent file is a perfect metaphor for a lot of early crypto engineering.

Then there's the human layer: theft and loss. NFTs are bearer-like assets. If someone steals the private key controlling your wallet, the NFT can be transferred away, and the chain will treat that transfer as valid. That's not a bug; it's what "ownership" means in a bearer system. Consumer-friendly NFT applications often reintroduce layers of account recovery, customer support, and sometimes even the ability to freeze or reissue tokens. Purists complain. Users, when it happens to them, tend to appreciate not being told "sorry, code is law."

One of the most underappreciated NFT use cases is boring corporate recordkeeping dressed as a collectible. Think warranties and authenticity certificates. Luxury goods have a counterfeiting problem, and authenticity is basically a battle over provenance. An NFT attached to a specific handbag, watch, or collectible can provide a verifiable chain of custody from manufacturer to retailer to buyer. But again, the token is only as good as the bridge between token and object. If anyone can attach an NFT to a fake watch, the ledger will faithfully preserve a fake history. Good systems require trusted issuance, tamper-resistant physical tags, and processes for disputes and repairs. The NFT is the public record; the rest is governance.

That word "governance" keeps showing up because NFTs are not only about uniqueness; they are about *coordination*. A ticketing issuer needs rules for resales and cancellations. A university needs rules for credential revocation (yes, it happens). A brand needs rules for transferring warranties. Identity systems need rules for privacy and recovery. The NFT itself is the vessel; the rules are the product.

There is also a quiet financialization that happens when receipts become tradable. A tradable ticket can become a speculative instrument. A tradable membership pass can become a status asset. A tradable in-game item can become a market. The moment you make a credential transferable and liquid, you invite market behavior: hoarding, arbitrage, manipulation, and price discovery that might not align with the issuer's goals. Tokenization doesn't merely digitize an object; it often turns it into a micro-market.

That's why the best NFT designs are honest about what is being made liquid and why. Sometimes liquidity is the point: a collectible that can be sold easily is more attractive to collectors. Sometimes liquidity is a side effect: a conference badge that can be resold needs rules so it doesn't become a scalper's paradise. Sometimes liquidity is dangerous: a credential that should not be sold (say, a professional license) should not be implemented as a freely transferable token. Technical capability is not the same as appropriate design.

In practice, NFTs are less about pictures and more about *serialization*. They give digital systems a native way to say: "This one is number 7, and it belongs to that person, and we can prove the chain of custody." Serialization is the backbone of titles, certificates, and collectibles. The internet didn't kill serialization; it just made it harder to enforce without centralized platforms. NFTs are one attempt to make serialization portable.

So the most useful way to think about NFTs is as a new kind of database entry with unusually strong properties: global uniqueness, transparent ownership history, and easy transfer. Those properties are powerful when you want a durable reference to a specific item or right. They are also insufficient on their own. If the NFT needs to map to a real-world right, you need a legal and operational bridge. If the NFT needs to protect consumers, you need recovery and dispute processes. If the NFT needs to preserve content, you need durable storage. The token is the handle; the system around it determines whether the handle is attached to something solid.

The cartoon ape era will be remembered for its excesses, but it did one important thing: it taught millions of people how to hold a unique digital asset and how to verify its provenance without asking a platform for permission. That is not a trivial cultural shift. Once people understand the idea of a scarce digital receipt, it becomes much easier to imagine credentials that travel with you, tickets that can't be forged, authenticity records that survive a brand's database migration, and identity proofs that don't require a password reset email.

In other words, NFTs are not the Mona Lisa of finance. They are the notary stamp. And as anyone who has ever tried to buy a home, board a plane with the wrong name on the ticket, or prove a credential across borders can tell you: notary stamps have a way of becoming interesting right when you need them most.

Chapter 4

Market Infrastructure and Operations

If blockchain is the engine of the new financial system, market infrastructure is the transmission that actually moves the car. We often obsess over the price of digital assets while ignoring the pipes that move them. This chapter lifts the hood on the 'plumbing' of the crypto economy. From the anxiety-inducing responsibility of holding your own private keys to the complex machinery of institutional vaults and decentralized exchanges, we will explore how value actually moves, settles, and stays safe in a world designed to function without traditional middlemen.

4.1 Self-custody and key management

Owning a digital asset can feel like owning a share in a company—until the day it behaves like a $100 bill that slipped behind the couch. The technology is new, but the risk is ancient: *possession matters*. In self-custody, there is no help desk that can "restore access" because access isn't an account privilege granted by an institution. It's the practical ability to produce a valid cryptographic signature. If you can sign, you can move. If you can't, you can't. Everything else is user interface.

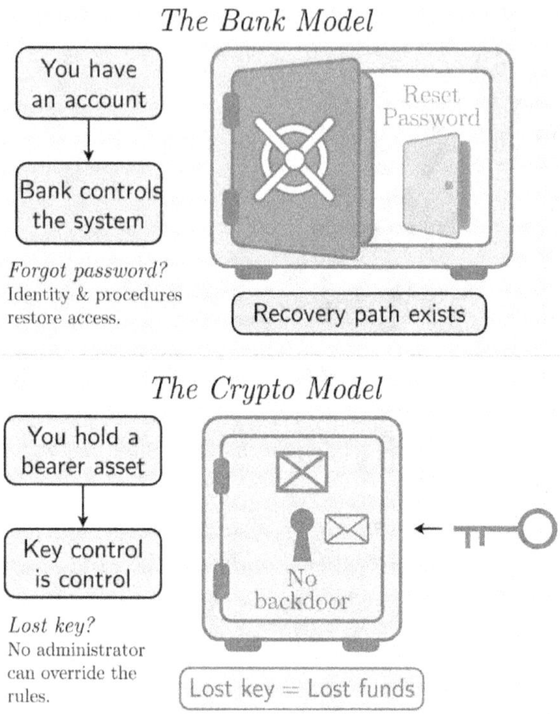

That difference—*account* versus *bearer asset*—is the entire movie. In the account world, a bank (or broker) keeps the definitive ledger and then grants you permissions. Your password is just a way to prove you're you. In the bearer world, the ledger is public and the permission is mathematical: whoever can produce the right signature can authorize the move. The password-reset door disappears because there is no security administrator above the system. There's just the system.

To make this concrete, separate two things that people casually mash together as "having crypto": an *address* and a *key*. A public address is like a glass mailbox bolted to the sidewalk. Everyone can see what's inside it (balances, tokens, past deposits), and anyone can send something to it. The private key is the mailbox key that lets you open the slot and take things out. You can paint your mailbox any color you want, check it from any phone you want, and tell your friends the address without fear. The danger lives in the key: if someone else gets it, they don't need your name, your ID, or your permission. They can just empty the mailbox.

Most modern wallets don't hand you a raw private key. They hand you something that looks almost insultingly simple: a list of 12 or 24 ordinary words. That "seed phrase" (also called a recovery phrase) is the master secret from which many private keys can be generated. The words aren't magic; they're packaging. Humans copy words more reliably than hexadecimal strings. But the security meaning is blunt: whoever has that phrase can recreate the keys and sign transactions. Write it on a sticky note, photograph it, email it to yourself, store it in a cloud notes app—and you have effectively made a bearer asset easier to steal than cash.

Self-custody is often marketed as freedom, and it is. It's also operational work, and the work doesn't come with a warning label loud enough for most first-time users. The easiest mental model is not "investing" but "handling valuables." Imagine a small velvet pouch that contains a key to a vault that can be accessed from anywhere in the world at any hour. You can carry it yourself, lock it in a home safe, or entrust it to someone else. In any case, the pouch is the product. The tokens are just what the key can move.

Wallets are not where assets live

A common beginner mistake is to think a wallet *stores* coins the way a leather wallet stores cash. On most blockchains, the assets don't sit in your app. They sit in the ledger's record of balances and ownership. The wallet stores (or helps you use) the keys needed to sign. That distinction matters because it changes what "backup" means. You are not backing up coins. You are backing up the ability to prove authorization.

This is why a broken phone is not necessarily a disaster and, paradoxically, why a perfectly functioning phone can be. If you have a properly recorded seed phrase, a lost phone is an inconvenience: you restore the wallet on a new device and move on. If you do *not* have the seed phrase, a lost phone can become permanent loss. Meanwhile, a working phone with a wallet on it is a live target. Malware doesn't have to "steal coins." It just has to trick you into signing something you didn't mean to sign, or extract a secret once and sign later.

The result is an odd inversion of normal consumer technology instincts. With most apps, convenience features like cloud backup and cross-device sync are safety nets. In self-custody, those same features can become trapdoors. "Sync my secrets to every device I own" is a great idea for photo libraries. It's a questionable idea for bearer keys.

Hot wallets, cold wallets, and the real meaning of "offline"

People talk about *hot* and *cold* wallets as if it's a temperature, but it's really an attack surface. A hot wallet is connected (directly or indirectly) to the internet and can sign transactions in an environment that also runs a thousand other processes. Cold storage keeps the signing keys offline so that even if your laptop is riddled with spyware, the keys remain out of reach.

Software wallets on phones and laptops are typically hot wallets. They're fast, convenient, and perfect for small balances or everyday activity. They're also operating in the same ecosystem as your web browser, your email, your downloads folder, and every link a friend sends you. When a hot wallet fails, it often fails not because the cryptography broke, but because the human did what humans do: clicked, installed, approved, trusted.

Hardware wallets are commonly used for cold storage. They are small devices designed so the private key never leaves the device. The transaction gets constructed on your computer, then the hardware wallet signs it internally, then it hands back the signature. That extra step is not ceremony; it's separation of duties. Your internet-connected computer becomes the "drafting table." The hardware wallet becomes the "signing room." Even if the drafting table is compromised, the attacker still needs you to confirm what's being signed.

Cold storage is not a single setting you toggle. It's a behavior. If the seed phrase is typed into a computer "just once" for convenience, it's no longer cold. If it's photographed and sits in an automatically backed-up camera roll, it's no longer cold. If it's stored in a password manager that syncs to the cloud, it might be convenient, but you have shifted your risk from local theft to account compromise and provider exposure. Sometimes that trade is rational. The point is to make it consciously, not accidentally.

The signature: your one true "authorization"

Traditional finance is authorization by institution. A wire transfer happens because a bank accepts your instruction, runs it through policies, and sends a message onward. Blockchain transfers are authorization by signature. When you sign, you are not asking permission. You are *creating a valid command* that the network will accept.

That power creates a peculiar kind of finality. A bank can reverse a transaction under certain conditions, freeze an account, or pause suspicious activity. A blockchain cannot "get suspicious" in a human sense. It executes rules. If you sign a transaction that sends assets to an attacker, the network will faithfully, efficiently, and impersonally do exactly what you asked. The fraud, if it occurs, is often upstream: deception in the user interface, phishing, malicious approvals, or coercion.

That's why key management is not just about hiding a secret. It's also about controlling *when* and *why* signatures happen. The key isn't a crown jewel you lock away and forget. It is a tool you use. And tools cause accidents.

Seed phrase hygiene: availability is as important as secrecy

Security conversations fixate on confidentiality: "Don't let anyone see it." For bearer assets, availability is equally existential: "Don't lose it." A perfect secret that is perfectly lost is still a loss.

A good way to think about the seed phrase is that it is both a *master key* and a *spare key*. It grants total control, and it's your fallback if the primary device fails. This dual role creates the central tension of self-custody: you want the seed phrase stored in a way that is accessible to you under stress (travel, disaster, device failure) but inaccessible to everyone else.

Paper is underrated here because it is stubborn. It doesn't get hacked. It doesn't silently sync. But it burns, it gets wet, it fades, and it can be thrown away by a well-meaning housemate doing spring cleaning. Metal backups (seed words stamped on steel) exist for a reason: they are boring, and boredom is a security feature.

The biggest enemy is convenience that leaves a trace. A photo of seed words is instantly copyable. A note file is instantly searchable. A screenshot is instantly uploaded to multiple devices if you have backups enabled. If you remember one rule, make it this: *never let your seed phrase exist as a digital file in a general-purpose device.* The moment it does, you're not just protecting against a burglar. You're protecting against every piece of software that can read that file, every service that can sync it, every account that can be phished, and every breach you'll only hear about six months later.

Wallet hygiene is mostly boring habits

The dramatic stories are about spectacular hacks. The everyday losses are more mundane: a seed phrase typed into a fake website, an approval signed without reading, a wallet connected to a malicious app, a laptop infected by a "free" cracked program. Wallet hygiene is less like building a fortress and more like not leaving your keys in the car.

A few habits do an outsized amount of work:

- *Treat your wallet like a spending account.* If you keep all your assets in the same hot wallet you use to connect to random decentralized apps, you are effectively carrying your net worth in your pocket. Many experienced users separate wallets by purpose: a "daily driver" for small amounts and experimentation, and a "vault" that rarely signs anything.
- *Read what you sign.* Wallet software often asks you to confirm amounts and addresses, but decentralized applications can also request broader permissions ("approvals") that let a smart contract move tokens later. This is the on-chain equivalent of giving someone a key rather than paying them once. Approvals are not inherently bad—they enable smooth trading and lending—but they are a favorite tool of scammers because they don't look like a theft in the moment. The theft comes later, when the approval is used.
- *Verify destinations out-of-band.* Clipboard malware exists. It swaps the address you copied with the attacker's address. The screen looks normal, your paste looks normal, and the transaction looks irreversible. For meaningful transfers, compare the first and last characters of the address, and if you can, confirm via a second channel (a hardware wallet screen, a known contact method, or a previously verified address book).
- *Keep your device clean.* This sounds like generic advice until you remember what a wallet does: it authorizes irreversible transfers. Don't install questionable browser extensions. Don't run pirated software. Don't treat your primary signing device like a sandbox. If you want to experiment, use a separate browser profile or a separate device with small balances.

- *Assume every QR code can lie.* QR codes are just an encoding method, not a trust signal. A QR code on a flyer can point to a phishing site. A QR code in a chat can encode a transaction request, not just an address. The convenience is real; so is the opportunity for misdirection.

If this starts to feel paranoid, that's because bearer assets turn ordinary consumer behavior into financial operations. The same phone that streams movies is also capable of signing away a portfolio in seconds. In traditional finance, the system is designed so that consumer-grade devices can be used safely because institutions absorb much of the operational risk. In self-custody, you inherit that risk. The device didn't change. The consequences did.

Key management as a lifecycle, not a one-time setup

Many people treat wallet setup like installing an app: do it once, forget it forever. Key management is closer to home maintenance. The moment you generate a seed phrase, you've started a lifecycle with predictable events: you will change phones, you will travel, you will update operating systems, you may have a theft or a fire, you may want heirs to access assets, you may decide you no longer trust a device. The question is not whether life will interrupt your perfect setup. The question is whether your setup was designed for interruption.

A practical self-custody lifecycle looks like this:

- *Generation.* Create keys in a trustworthy environment. New, reputable hardware wallets and well-audited software wallets reduce risk, but the real vulnerability is often the human environment: cameras, screensharing, people nearby, and any temptation to "just save it digitally."
- *Storage.* Store the seed phrase so that it survives time. Paper in a desk drawer is not a plan; it's a hope. If you wouldn't

store a stack of cash that way, don't store the master key to a bearer wallet that way. Consider theft, fire, flood, and simple misplacement.

- *Use.* Decide how often keys will be used, and in what context. The less frequently a vault wallet signs, the smaller its exposure. When you do use it, slow down. Speed is the enemy of verification.
- *Backup and recovery testing.* A backup you've never tested is a rumor. Recovery testing doesn't mean moving money around; it can mean restoring a wallet in a controlled way and verifying you can see the expected addresses without broadcasting transactions. The time to learn you copied a word wrong is not the day you lose your phone.
- *Compromise response.* If you suspect the seed phrase is exposed, treat it like a stolen house key. You don't argue about whether the thief will come back. You rekey. In crypto terms, that means moving assets to a new wallet with a new seed phrase. The speed requirement here is why having a plan matters. Under stress, people make mistakes that compound losses.

The underlying theme is maturity: self-custody is not about memorizing a clever trick. It's about building a small operational system that can tolerate human life.

The hidden trap: identity scams and "support"

One of the strangest aspects of crypto is that the biggest security hole is often social, not technical. There is an entire ecosystem of scams that exploit the psychology of self-custody: the panic of a stuck transaction, the confusion of unfamiliar interfaces, the fear of losing funds. Attackers don't need to crack cryptography when they can simply persuade you to hand over the seed phrase.

The easiest bait is "customer support." A malicious account impersonates a wallet provider, an exchange, or a popular app. They offer to help. They ask you to "verify" your wallet by entering your seed phrase into a form. Or they ask you to install remote access software. Or they send you a link to "synchronize" or "validate" your wallet. The pitch is always the same: do something that feels like account recovery. But self-custody has no account recovery. There is only key disclosure.

A reliable rule: *no legitimate support agent will ever need your seed phrase*. Not to troubleshoot. Not to verify. Not to "reset." If someone asks for it, the conversation is over.

Multisig and MPC: reducing single-point-of-failure risk

Self-custody doesn't have to mean one key, one device, one moment of catastrophe. Two families of tools try to solve the same problem: reducing dependence on a single secret.

A *multisignature* (multisig) wallet requires more than one key to authorize a transaction—for example, 2-of-3 keys. You might keep one key on a hardware wallet at home, one in a safe deposit box, and one with a trusted person or stored in a separate secure location. The beauty is that losing one key doesn't mean losing funds, and an attacker who steals one key can't drain the wallet. The trade-off is complexity: you must manage multiple devices and be careful not to accidentally concentrate the keys in the same place (or the same cloud account).

Multi-Party Computation (MPC) is a different approach. Instead of multiple independent keys, MPC splits signing authority into "shares" distributed across devices or parties. The signature is produced collaboratively, and no single party ever holds the full key in one place. From the user's point of view, MPC can feel more seamless than multisig. From a risk perspective, it can reduce the chance that a single stolen device leads to loss. The flip side is that implementations vary widely, and some MPC setups introduce a service provider into your trust model. It may still be self-custody in the sense that no one party can move funds alone, but it is not the same as having all secrets entirely under your personal control.

Both approaches are best understood as *operational design*, not magic. They force an attacker to compromise multiple things. They also force you to plan your own recovery story. If your setup depends on three devices and one of them is lost, what happens? Who can help, and what exactly do they have the power to do?

A practical checklist that doesn't pretend you're a cybersecurity engineer

Self-custody advice often oscillates between two unhelpful extremes: casual ("just write down the words") and intense ("build a Faraday cage and bury a titanium plate at a secret location"). Most people need something in the middle: a disciplined routine that respects the stakes without turning life into a spy novel.

A workable baseline looks like this:

- *Decide your security tier by balance, not by enthusiasm.* Use a hot wallet for small, active amounts. Use a cold or hardware wallet for meaningful holdings. If the value would hurt to lose, it deserves a higher tier.
- *Create the seed phrase privately and record it offline.* No screenshots. No cloud notes. No emailing yourself. If you must transcribe, do it carefully and double-check every word.
- *Store the backup like a physical valuable.* Use a secure location. Consider environmental risks (fire/water). Consider theft. Consider whether a single point of storage creates a single point of failure.
- *Test recovery once, calmly.* Restoring a wallet for the first time while stressed is how people make expensive errors. Do a con-

trolled restore and verify that you can reproduce the correct accounts.

- *Minimize approvals and periodically review them.* Think of token approvals as keys you've handed out. If you no longer use an application, consider revoking its permissions using reputable tools and official links.
- *Slow down when moving size.* For large transfers, do a small test first. Verify addresses. Verify on a trusted screen when possible. Treat it like wiring money, not like sending a text.

This may sound like a lot. It is less work than recovering from a loss that cannot be appealed.

The inheritance problem: keys don't care about your will

Traditional finance has a built-in interface to death: beneficiaries, transfer-on-death registrations, probate procedures, and institutions that recognize legal authority. Bearer assets are indifferent. A blockchain will not read a will. It will not respond to a court order. It will respond to signatures.

That creates a responsibility most people don't consider until it's too late: if you self-custody, you have created an asset that may be *unrecoverable by your family* unless you design a path for them. The goal is not to weaken security by leaving secrets lying around. The goal is to create a controlled process that works in the real world.

Some people document where the seed phrase is stored and how to use it, sealed and kept with estate documents. Others use multisig so that heirs and trusted parties can collaborate without any one person holding unilateral power. There is no universal best practice because family situations vary. The hard truth is that "I'll remember it" is not a plan, and "my spouse will figure it out" is often a tragedy in slow motion.

Why self-custody feels harder than it "should"

If self-custody sometimes feels unforgiving, it's because it is closer to the settlement layer than most consumer finance products allow you to touch. In a bank, you are insulated from the machinery. You can make mistakes and still have procedural guardrails. In self-custody, you are closer to the machinery, and the guardrails are mostly the ones you build: device choices, backup discipline, and signing habits.

The good news is that the basic competence threshold is not unattainable. You don't need to understand elliptic curves to understand that a seed phrase is a master key. You don't need to read smart contract code to understand that approvals grant ongoing power. You don't need to become paranoid to become careful.

The most valuable mindset shift is simple: stop thinking of a wallet as an app, and start thinking of it as a *cash-and-keys workflow*. Once you do, the best practices stop sounding like superstition and start sounding like what they are: ordinary operational controls for an extraordinary kind of asset.

4.2 Institutional safekeeping: Custody models

A retail wallet can run on courage, common sense, and a piece of paper hidden in a kitchen drawer. Institutional custody cannot. A pension plan, an ETF sponsor, or a corporate treasury doesn't get to say, "I'm pretty sure the seed phrase is in my email somewhere." The job is not merely to prevent theft; it's to create a system where theft is difficult, errors are containable, and evidence exists when auditors, regulators, and clients ask the most uncomfortable question in finance: *prove it*.

Custody, in institutional terms, is a promise with teeth. The custodian is expected to keep assets safe, keep records straight, process instructions correctly, and survive bad days: employee turnover, cyber incidents, market stress, and even the custodian's own financial distress. The technical heart of digital-asset custody is key control, but the real body is operations: governance, segregation, reconciliations, access controls, and the legal structure that decides whose assets are whose when something breaks.

The word ``custody" means different things depending on who is paying attention

In casual crypto conversation, "custody" often means "who has the keys." In regulated finance, custody is bigger: it's a blend of safekeeping, recordkeeping, and responsibility. A qualified custodian is not just a vault; it's a regulated entity with duties and scrutiny. This is why banks and trust companies get so much attention in U.S. crypto conversations: regulators have decades of muscle memory for what bank custody is supposed to look like, and certain regulators have explicitly treated crypto custody as an extension of traditional custody activities, with the twist that safeguarding may involve holding cryptographic keys.

That framing matters because it changes the standard of care. A retail user can be forgiven for making an imperfect trade-off between convenience and security. An institution cannot. It must justify its controls the way it would justify controls over cash, securities, or customer data: documented policies, tested procedures, clear roles, and a paper trail that survives cross-examination.

The first fork in the road: trading venue custody versus dedicated custody

Many institutions meet digital assets through exchanges. Exchanges are useful because they combine trading and custody into one seamless surface: deposit in, trade, withdraw out. That convenience is also the risk: if the venue is holding assets while also running a trading business, it becomes a single point of failure across multiple dimensions—operational, legal, and sometimes ethical.

Dedicated custodians try to separate the safekeeping function from the trading function, the same way traditional markets separate custody from brokerage and exchange functions. In the cleanest version of this model, an institution uses:

- a trading venue (or broker) for execution,
- a custodian for safekeeping,
- and a controlled movement process between them.

The goal is not to make trading harder. The goal is to stop "the place where you click buy" from also being "the place where your entire inventory quietly lives."

That separation is operationally annoying in the way seatbelts are annoying: it adds steps, it slows you down, and it is absolutely the point.

Commingled versus segregated: the difference between ``trust me'' and ``show me''

The phrase *asset segregation* sounds like back-office jargon until you translate it into plain English: "If the custodian fails, are those assets clearly mine?" Segregation is as much about bankruptcy and bookkeeping as it is about technology.

Commingled (or omnibus) custody means many customers' assets are held together in shared addresses or accounts, with internal ledgers tracking who owns what. This is how much of traditional finance works, and it can be safe *when the institution is well-run and well-regulated*. It can also be disastrous when internal books are sloppy, when controls are weak, or when leadership treats client assets as a financing tool.

Segregated custody aims to give each customer a cleaner boundary: distinct wallets or on-chain addresses, distinct sub-accounts, and operational processes that keep customer property separate from the custodian's own assets. Done well, segregation does two things:

1. It reduces operational ambiguity—reconciliations are clearer because the on-chain picture can map more directly to the customer ledger.
2. It strengthens the customer's claim in insolvency scenarios by making it harder to argue that "everything was mixed together anyway."

Segregation is not free. It can increase transaction costs, increase wallet-management complexity, and create a larger "surface area" of addresses to monitor. Institutions pay that price because the alternative is often a risk they cannot explain to an investment committee without sweating.

Custody design choice	*Commingled / omnibus*	*Segregated (per client / per account)*
On-chain structure	Shared wallets; internal ledger allocates ownership	Separate wallets or clearly partitioned addresses per client
Operational simplicity	Easier to operate at scale; fewer wallets	More operational overhead; more addresses and policies
Proof and reconciliation	Heavy reliance on custodian's records; harder to validate externally	Clearer mapping between on-chain holdings and customer positions
Insolvency clarity	Can become legally and operationally messy if records are weak	Typically stronger customer-asset separability when implemented rigorously
Common failure mode	Ledger mismatch, hidden liabilities, or misuse of pooled assets	Misconfiguration, address management errors, or higher operational complexity

The lesson isn't that commingling is always bad and segregation is always good. The lesson is that commingling turns custody into a *trust-heavy* model. Segregation shifts the burden toward *verifiability*. Institutions tend to prefer verifiability, especially when headlines have demonstrated what happens when "trust me" meets leverage.

Qualified custody: less about gadgets, more about governance

When institutions ask, "Is this a qualified custodian?" they are really asking, "Is there a mature control environment and a regulator who can compel the ugly work?" The sophisticated part of custody is not the hardware; it's the discipline.

A mature custody environment looks like a choreography of constraints:

- *Who* can initiate movement?
- *Who* can approve it?
- *What* policies must be satisfied (limits, whitelists, time delays)?
- *How* is it recorded and reconciled?
- *What* happens if something goes wrong at 2 a.m. on a holiday weekend?

In other words, institutional custody borrows the logic of traditional cash controls: dual control, separation of duties, audit logs, and limited authority. In the digital-asset world, these controls are enforced through key management and transaction workflows rather than through bank wires and manual sign-offs, but the spirit is identical: no single person should be able to move client assets unobserved.

Key storage: HSMs, vaults, and the myth of the magic box

Institutions rarely use consumer hardware wallets for serious balances. Not because hardware wallets are inherently bad, but because institutions need enterprise features: centralized policy enforcement, role-based access, tamper resistance, secure backups, and auditability. This is where hardware security modules (HSMs) and specialized custody platforms enter the story.

An HSM is a hardened device designed to generate, store, and use cryptographic keys inside a tamper-resistant boundary. The promise is simple: keys can be used for signing, but never exposed as raw material. The value proposition is not just that the keys are hard to steal. It's that the signing process can be wrapped in controls: approvals, thresholds, and monitored access.

Still, it's dangerous to treat HSMs as magic. An institution can buy expensive hardware and still lose money through:

- misconfigured policies,
- compromised operator credentials,
- poor segregation of environments (e.g., dev and production),
- or flawed transaction validation.

A vault can be world-class and the teller can still hand cash to the wrong person. In custody, the transaction is the moment of truth. Everything else is preparation.

MPC and multisig in institutions: less heroism, more risk math

Institutions like tools that turn catastrophic risks into manageable ones. Two techniques do this particularly well: multisignature wallets and multi-party computation (MPC).

Multisig splits authority across multiple keys so that no single key can move assets. A common arrangement might require 2-of-3 approvals. This makes theft harder because an attacker must compromise multiple keys. It also makes insider fraud harder because one rogue employee cannot act alone.

MPC takes a different route: instead of multiple independent keys, it splits a key into shares so that signing requires collaboration among parties or systems. The practical advantage is flexibility: MPC can be integrated into enterprise workflows with less friction than on-chain multisig in some contexts, depending on the chain and asset type. The practical risk is that MPC setups vary widely; the trust model depends on who holds the shares, who operates the infrastructure, and how recovery works under stress.

Institutions choose between these approaches the way engineers choose between bridges: not by aesthetics, but by load and failure modes. The aim is to remove single points of failure, because "single point of failure" is finance-speak for "future headline."

Asset movement controls: the institutional equivalent of ``are you sure?"

Retail wallets often rely on a single confirmation screen. Institutions build a gauntlet.

A serious custody workflow typically includes:

- *Whitelisting*: assets can only be sent to pre-approved addresses.
- *Transaction limits*: thresholds that require escalating approval.
- *Time delays*: large withdrawals may have enforced cooling-off periods.
- *Dual control*: one person initiates, another approves.
- *Out-of-band verification*: approvals may require separate devices or channels.

These controls feel slow because they are designed to defeat the two enemies of institutional finance: haste and silence. Haste causes fat-finger errors. Silence enables fraud. A well-designed custody workflow makes both difficult.

Even the act of confirming an address changes at institutional scale. A retail user might compare the first and last characters. A custodian may rely on address books, beneficiary management, and policies that stop new destinations from being used until they pass review. It's not because institutions are smarter; it's because institutions are sued when they are wrong.

Proof of reserves: a helpful idea that can become theater

After high-profile failures in the crypto industry, the phrase *proof of reserves* gained popularity: show cryptographic evidence that assets exist. It's an intuitively appealing idea because blockchains allow public verification of balances.

But reserves are not the whole story. A custodian can show assets and still be insolvent if it has hidden liabilities. It can also show assets that are encumbered or pledged elsewhere. Proof of reserves can be a useful transparency tool, especially when paired with robust audits and liability disclosures, but it can also become performance art: a snapshot that reassures without truly informing.

Institutions should treat proof of reserves like a bank statement from a stranger: interesting, but not sufficient. The real question is not "does this address have coins today?" but "is the control environment strong enough that the coins will be there tomorrow, and are they legally and operationally dedicated to the right owners?"

Sub-custody chains: the risk you inherit without noticing

Custody is often a supply chain. A bank might outsource technology to a specialist. A custodian might rely on a sub-custodian for certain assets. A platform might depend on an MPC provider. This is normal in modern finance; very few institutions build everything themselves.

The risk is that outsourcing can quietly multiply the number of entities that matter. You may believe you are relying on a single custodian, but your operational reality might involve:

- the custodian,
- a technology vendor,
- a cloud provider,
- a sub-custodian or settlement agent,
- and an incident-response firm on retainer.

Each link can be well-managed; each link can also fail. Institutions should map these dependencies the way they map counterparty exposures. If there is a breach, a service outage, or a legal dispute, who has what power, and how fast can assets be moved to safety?

Accounting and disclosure: custody can change what a balance sheet looks like

Custody sounds operational until accounting gets involved, and then it becomes existential. If a platform safeguards assets for customers, what must it recognize on its own balance sheet? What risks must it disclose? How should it describe who controls the keys and what happens in insolvency?

These questions have been contentious enough that they've produced specific staff guidance and later revisions in the U.S. The important point for an operator is not the footnote number; it's the message: regulators and auditors consider crypto safeguarding to have unique risk characteristics—technological, legal, and operational—and those characteristics can affect financial reporting and business models.

That creates a feedback loop. If custody treatment imposes significant balance sheet impact or disclosure burdens, some firms will avoid direct custody and instead partner with specialized custodians. Others will invest heavily to meet requirements. Market structure often follows accounting gravity.

So what does ``safe" custody look like in practice?

Safety in custody is rarely dramatic. It looks like processes that hold up under boredom:

- Daily reconciliations between on-chain positions and internal ledgers.
- Exception handling that is documented and reviewed.
- Regular key-ceremony procedures with witnesses and logs.
- Access reviews that remove permissions when employees change roles.
- Incident drills that assume failure will happen and measure response time.

The most telling sign of maturity is not that an institution claims it is secure. It is that it can explain, in plain language, what would happen in specific failure scenarios:

- "A signer laptop is compromised."
- "A key share is lost."
- "A vendor is breached."
- "A chain is congested and withdrawals stall."
- "A regulator demands records during a market crisis."

A good custodian treats those scenarios as operating conditions, not as hypotheticals.

Choosing a custody model is choosing which risks you can live with

No custody design eliminates risk; it reallocates it. Keeping assets on an exchange can reduce friction but increases exposure to venue failure and commingling risk. Fully segregated qualified custody can reduce insolvency ambiguity and improve governance, but it can increase cost, complexity, and operational dependency on workflow tooling. MPC and multisig reduce single-key catastrophe, but they introduce their own recovery and governance requirements.

The institutional mindset is not "What is the perfect system?" It's "What is the system whose failure modes we can understand, test, insure, and explain?" In markets, mystery is risk. Custody models are ultimately a way of refusing mystery—by turning key control into policy, policy into process, and process into evidence.

4.3 Trading venues: CEX, DEX, and Liquidity

Buying a token is rarely a single act. It's a small choreography of trust: trust in a venue, trust in a price, trust that the asset you think you're buying is the asset you'll be able to sell. Traditional markets hide most of that choreography behind familiar nouns—*broker*, *ex-*

change, clearing. Digital-asset markets put the machinery on the table. Sometimes it's wrapped in a slick app; sometimes it's exposed as raw smart-contract calls. Either way, the venue is not a detail. It is the product.

Centralized exchanges (CEXs) feel like the internet's version of a brokerage account. You sign up, complete identity checks, deposit dollars (or another asset), click buy, and see a balance change instantly. The trade is fast because it is mostly internal: the venue updates a database, matches orders, and promises that the ledger will line up when you withdraw. The promise might be honored brilliantly. It might also be tested at the worst possible moment—a market crash, a bank run, an outage—when every customer suddenly wants the same thing: their assets, right now.

Decentralized exchanges (DEXs) feel like the opposite. There is often no account, no application form, no customer service queue. You connect a wallet and trade directly against a protocol: code deployed on a blockchain. The trade is slower because it rides on-chain settlement, and it is transparent because the state of the system is visible to anyone who looks. But transparency does not mean simplicity. A DEX trade can be a small bundle of permissions, approvals, routing hops, and fees. With a CEX, you outsource complexity. With a DEX, you accept it, and then try to tame it.

What a venue actually does: price discovery and inventory transfer

Strip away the branding and the venue's job is twofold: *find a price* and *move inventory*. "Price discovery" is market jargon for the first part: the process by which a crowd of buyers and sellers reveals what something is worth right now. The second part is the plumbing: once the price is agreed, who ends up with what.

A CEX typically uses a *central limit order book* (CLOB), the classic mechanism of stocks and futures. Traders post bids (offers to buy) and asks (offers to sell). The highest bid and lowest ask form the visible market. If you want instant execution, you cross the spread:

you buy at the ask or sell at the bid. If you want a better price, you place a limit order and wait.

A DEX can use an order book too, but the dominant design in crypto has been the automated market maker (AMM): a pool-based system where prices are determined by a formula and the pool's current balance of assets. It's less like a traditional exchange floor and more like a vending machine that changes its price as its shelves empty.

Both designs can work. Both have sharp edges. The interesting differences show up when you ask a practical question: *What happens when I trade size?*

Liquidity: the invisible ingredient that determines whether you get a fair deal

Liquidity is one of those words that sounds abstract until you meet it in the wild. It means the ability to buy or sell without moving the price too much. In a liquid market, you can trade and the market barely notices. In an illiquid market, your trade *is* the market.

Liquidity has two faces:

Depth: how much can be traded near the current price. Deep markets can absorb large orders with minimal price movement.

Tightness: how small the gap is between the best bid and best ask. That gap—the *bid-ask spread*—is the market's cover charge. You pay it even when you think trading is "free."

CEXs often feel liquid because the interface is familiar and the matches are fast. Many large CEX order books are indeed deep for major assets. But liquidity is not a property of the website; it is a property of the *order book* and the *participants* on it. A token can trade on a big-name venue and still be thin. In stress, even deep books can turn to ice as market makers widen spreads or step away.

DEX liquidity is more legible. You can often see the pool sizes on-chain. That visibility is comforting, but it can also be sobering: many pools are small, fragmented across multiple DEXs and fee tiers, and sensitive to large orders. DEX liquidity can be abundant in aggregate and still painful at the point of execution if your trade is routed through a shallow pool.

The bid-ask spread: the fee that doesn't send you a receipt

People obsess over explicit fees and miss the larger cost: trading at a worse price than they expected. The spread is the simplest example. If the best bid is $99 and the best ask is $101, and you buy immediately, you pay $101. If you sell immediately, you receive $99. The $2 difference is not a fee charged by the venue in the way a commission is. It's a cost paid to liquidity providers and market makers for standing ready to trade.

A CEX often displays the spread clearly, but it can be psychologically hidden by the speed of execution: you click, you get filled, you move on. A DEX can hide it differently: you might be quoted a price that looks good, but the final execution incorporates *slippage* (the price moves against you as your trade changes the pool balance) plus network fees plus any routing hops.

The important habit is to treat the displayed price as an invitation, not a guarantee. The true cost of trading is the executed price relative to a fair reference price—and the difference between those two is where liquidity (or lack of it) makes its money.

How AMMs set prices: constant product and the art of unavoidable slippage

The canonical AMM design—popularized by Uniswap—uses a simple constraint: the product of the pool's two asset balances stays constant, often written as $x \cdot y = k$. If a pool holds x units of token A and y units of token B, the relationship pins down a curve of possible states. When you buy A from the pool, you remove some x, and you must add enough y to keep the product aligned (after fees). The pool is always willing to trade, but it demands a worse marginal price as you take more of one side. That worsening marginal price is slippage made mechanical.

This is a profound design choice. An order book is a crowd of discrete opinions: multiple sellers at multiple prices. An AMM is a single continuous rule: one pool, one curve, a price that responds immediately to inventory changes. It is elegant and brutally honest. If the pool is small relative to your trade, the curve punishes you.

That punishment is not a bug. It is a protection mechanism. Without it, a pool could be drained at stale prices. Instead, the AMM forces large traders to either accept slippage, split trades, route across multiple pools, or use specialized mechanisms. Liquidity becomes a tangible thing: a pool with more capital has a flatter curve and can offer better execution.

There is also a subtle social contract at work. AMM liquidity is provided by liquidity providers (LPs) who deposit assets into pools in exchange for fees. They are paid for enabling trades, but they bear risks—including *impermanent loss*, the opportunity cost of providing liquidity when prices move sharply. The DEX fee you pay is not just revenue for a company. It is compensation for other market participants taking the other side of your urgency.

DEX transparency: you can see the machine, but you can still lose a finger

The transparency of DEXs is real. The smart contracts are deployed on-chain. Trades are publicly observable. Pool balances can be monitored. Anyone can audit the flows, at least in principle.

But transparency changes the game for attackers and sophisticated traders too. If your transaction is visible before it is finalized, others can react. This is where the market structure becomes almost theatrical: transactions sit in a public waiting room, and other participants may try to insert their own transactions around yours. The term *MEV* (maximal extractable value) captures a family of behaviors where someone profits by reordering, inserting, or censoring transactions. The practical takeaway for ordinary traders is simpler: on a DEX, you are not just trading against a pool. You are trading in a world where your intention can be observed and exploited if you are careless with slippage settings and routing.

This is why DEX interfaces ask for a slippage tolerance. Set it too tight, and your trade fails, wasting fees and time. Set it too loose, and you've effectively signed a blank check that says, "Fill me at any price up to this limit." In volatile or thin markets, that is an invitation for bad outcomes.

A CEX trader is also exposed to adverse selection and fast traders, but the dynamics are different. The matching engine is internal and not publicly broadcast transaction-by-transaction. The CEX has discretion in how it sequences orders, how it handles partial fills, and how it manages outages. DEXs outsource discretion to protocol rules; CEXs internalize it as operational policy.

CEX convenience: you get a dashboard and a referee

The CEX experience is popular for a reason. It packages a chaotic market into a coherent product:

- fiat on-ramps (bank transfers, cards),
- familiar order types,
- customer support,
- and often a smoother tax and reporting experience.

It can also act as a referee. If a customer fat-fingers an order, the exchange might have policies, error handling, or at least a support channel. It may also run surveillance, block suspicious withdrawals, and implement compliance checks. Some of these features feel intrusive; many become appreciated the day something goes wrong.

The price of this convenience is counterparty exposure. When assets are held on a CEX, you are trusting the venue's custody model, internal controls, and solvency. Even if the venue is honest, operational failures happen: wallet systems go down, withdrawals get delayed, and banking partners can become bottlenecks. In extreme scenarios, customers discover that "available balance" is not the same thing as "immediately withdrawable."

That's why sophisticated institutions treat exchange balances like working inventory, not long-term storage. They may keep enough on-venue to trade efficiently and sweep excess to a custodian. This is not paranoia; it is inventory management.

DEX composability: your trade can be a Lego tower

DEXs don't just offer trading. They offer *composability*: the ability for different protocols to plug into one another. A DEX swap can be one step in a larger sequence: borrow against collateral, swap, provide liquidity, stake the LP token, and route yields elsewhere. This is where decentralized finance becomes more than trading; it becomes programmable capital.

Composability is powerful, but it creates a new kind of operational risk: stack risk. Each additional protocol is another set of smart contracts, parameters, governance decisions, and potential vulnerabilities. The trade can be transparent and still be fragile because the overall structure depends on many moving parts working exactly as expected.

The irony is that composability can make DEX trading better and worse at the same time. Better because routers can search across many pools to find the best price, effectively stitching together liquidity. Worse because routing introduces extra hops, extra fees, and extra points where something can fail or be manipulated.

Liquidity fragmentation: one asset, many pools, many prices

In traditional markets, fragmentation exists across exchanges, but it is moderated by market structure, regulation, and professional market makers. In crypto, fragmentation can be extreme. The same token may trade across:

- multiple CEXs,
- multiple DEXs,
- multiple pool types and fee tiers,
- and multiple chains (wrapped versions, bridged versions, synthetic versions).

Fragmentation affects price and execution. If liquidity is scattered, each venue can have a slightly different price. Arbitrageurs connect the venues by buying where it's cheap and selling where it's expensive, pushing prices back into alignment. This is healthy market function, but it also means that the quality of your execution depends on how well your venue is connected to the broader ecosystem.

CEXs often concentrate liquidity for major pairs, which can make execution cleaner. DEXs can be simultaneously fragmented and globally accessible, which makes routing quality a real differentiator. Two trades of the same size in the same token can have meaningfully different outcomes depending on which pools were used and what else was happening on-chain at the time.

Trust mechanisms: ``don't trust, verify" meets ``verify what, exactly?"

The crypto slogan "don't trust, verify" works beautifully as an attitude and imperfectly as a plan. Verification has costs. It requires data, tooling, and a clear question.

On a DEX, you can verify a lot: contract addresses, pool balances, fee parameters, and transaction history. You can also be fooled by lookalike tokens, malicious contracts, and interfaces that route you somewhere you didn't intend. Verification is possible, but it demands competence.

On a CEX, you cannot verify internal ledgers in the same way. You can evaluate governance, regulation, audits, and transparency reports. You can look for proof-of-reserves disclosures, which often combine on-chain evidence of asset control with cryptographic techniques (like Merkle trees) to let customers verify their liability inclusion in a snapshot. But these proofs are typically point-in-time and scope-limited, and they rarely capture the full balance sheet reality. They are directionally helpful and not a substitute for a robust control environment.

The practical stance for market participants is not maximal skepticism or maximal faith. It is calibrated reliance:

- Use CEXs for what they are good at (fast execution, fiat access, deep liquidity on majors).
- Use DEXs for what they are good at (transparent execution rules, self-custody compatibility, access to long-tail assets and programmable strategies).
- Treat both as venues with distinct failure modes rather than as moral categories.

Compliance and identity: why CEX flows feel bureaucratic and DEX flows feel wild

Centralized venues sit squarely in the world of AML/CFT (anti-money laundering / countering the financing of terrorism) obligations. That means customer identification, transaction monitoring, sanctions screening, and, increasingly, information-sharing expectations around transfers. The operational result is familiar friction: KYC forms, withdrawal holds, source-of-funds questions, and sometimes limits when interacting with self-custody addresses.

DEXs, by contrast, can be accessed with a wallet and an internet connection. That openness is part of their appeal, and also the reason they attract intense regulatory attention. Many compliance controls that are straightforward in an account-based system are difficult to implement in a protocol-driven environment without changing what the environment is.

For users, the trade-off is straightforward: CEXs offer a regulated on-ramp and guardrails at the cost of surveillance and permissioning; DEXs offer permissionless access at the cost of personal responsibility and fewer built-in protections. The venue choice quietly becomes a choice about what kind of adult you want to be in the financial system: the one with a concierge, or the one with a toolkit.

A trading reality check: the venue is part of your return

Investors like to think returns come from being right about an asset. In practice, returns are also shaped by where and how you trade. Two people can have the same market view and end up with different outcomes because one paid wide spreads, got hit by slippage, or couldn't exit during volatility.

A few practical questions separate thoughtful trading from wishful trading:

- *Where is the liquidity deepest for this asset, right now?*
- *What is my total cost: fees, spread, slippage, and network costs?*
- *What happens if the venue goes down or the chain congests?*
- *Am I holding assets on-venue longer than necessary for execution?*

These questions are not glamorous. They are the difference between a market participant and a market donor.

The most honest summary of CEX versus DEX is not that one is "safe" and the other is "risky." Both can be safe; both can be risky. A CEX concentrates trust into an institution and its controls. A DEX distributes trust into code, public state, and the user's own operational competence. The convenience of one and the transparency of the other are not marketing slogans. They are different ways of paying the same bill: the bill for participating in a market that does not owe you a good outcome.

4.4 Moving value: Settlement and bridges

A crypto transfer can feel like a magic trick: tap "send," watch a hash appear, refresh a screen, and there it is—value on the other side. The illusion is speed. The reality is that you've started a race between *visibility* and *irreversibility*. The transaction becomes visible quickly; it becomes truly difficult to undo more slowly. That gap is where operational risk lives, and it's why serious market participants obsess over a word most retail users never say out loud: *settlement*.

Settlement is the moment an obligation becomes final. In markets, it is the difference between "we agreed" and "it's done." Everything else—quotes, confirmations, cheerful green checkmarks—is prelude.

Finality: when ``confirmed" becomes ``can't be unwound"

Finality is a slippery concept because different systems mean different things by it. In traditional payment and securities systems, finality is largely a legal and institutional guarantee. A central bank or settlement system declares a transfer final under its rulebook, and that declaration carries weight because courts, regulators, and supervisors recognize it. The system can still reverse things in extraordinary circumstances, but reversals are *exceptions* executed by institutional authority.

Blockchains invert the logic. There is no administrator with a master switch. Finality is a property of consensus: a transaction becomes increasingly hard to reverse as more blocks build on top of it, or as the system reaches a consensus checkpoint that would be economically catastrophic to roll back. This is not legal finality; it is *crypto-economic* finality. The network doesn't promise it won't change its mind. It makes changing its mind very expensive.

That difference explains a common confusion. People talk about "confirmations" as if they are finality. A confirmation is a receipt that a block producer included your transaction. Finality is the point at which rewriting history would require so much coordination, stake, or bribery that rational actors won't attempt it. Confirmations are about being seen. Finality is about being stuck.

Probabilistic versus economic finality: why time is part of settlement

In some blockchains, finality is probabilistic: the deeper a transaction is buried under subsequent blocks, the less likely it is to be reversed. There is no single line in the sand. There is only a risk curve that approaches zero but never quite touches it.

Other systems provide a stronger notion of finality through explicit finalization mechanisms. Ethereum's proof-of-stake design, for instance, uses checkpoint voting by validators. When enough validators (a supermajority) vote in favor of a checkpoint, blocks become *finalized* in a technical sense: reverting them would require behavior that is punishable by severe economic penalties. This doesn't make reversals metaphysically impossible. It makes them economically self-destructive at scale.

The market implication is not that one chain is "good" and another is "bad." It's that settlement has a time dimension. If you run a venue, a broker, a payment processor, or a treasury desk, you end up deciding thresholds:

- How many confirmations are enough before crediting a deposit?
- What constitutes finality for our risk appetite?
- What happens during congestion or reorg events?

Those policies are not academic. They are the difference between smooth operations and an incident report with screenshots.

Settlement is a risk control, not a user interface feature

User interfaces are optimists. They show progress bars, spinners, and confidence. Operations teams are pessimists. They ask: *what could go wrong in the window between initiation and finality?*

That window is where the unpleasant but ordinary things happen:

- a transaction sits unconfirmed because fees were set too low,
- a chain becomes congested and confirmation times expand dramatically,
- a block reorganization temporarily removes a transaction from the canonical history,
- or a venue credits a deposit too early and becomes exposed to reversal risk.

In traditional finance, these are often absorbed by institutions and smoothed over. In crypto markets, they show up in customer experiences: delayed withdrawals, "pending" transfers, and sudden changes in required confirmations when volatility spikes.

A mature way to think about this is that settlement is not one event but a series of states: *broadcast, included, confirmed, finalized.* Different businesses choose different cutoffs. The cutoffs are where risk is priced.

Delivery-versus-payment: the old idea that keeps haunting new systems

Traditional market infrastructure has spent decades trying to eliminate *principal risk*—the risk that you deliver what you owe but don't receive what you're due. The classic solution is delivery-versus-payment (DvP): make the delivery of the asset conditional on payment, so neither side can end up naked.

Crypto systems sometimes achieve something like DvP inside a single transaction: swaps can be atomic, meaning they either happen entirely or not at all. That's a genuine innovation. But once you move across institutions, chains, or off-chain money systems, principal risk creeps back in.

A striking example is fiat settlement against on-chain assets. You can swap token A for token B in one atomic transaction on a DEX, but you cannot atomically swap token A for a bank wire in most real-world setups. That gap is where brokers, stablecoins, escrow arrangements, and payment rails re-enter the story. The future is often described as "on-chain everything." The present is a patchwork.

Bridges: interoperability's most expensive shortcut

If settlement is about finality on one ledger, bridging is about moving value between ledgers that do not share a native language. Blockchains are often called networks, but they are more like sovereign islands with their own rules. A bridge is the ferry service. It's also, historically, where the storms hit hardest.

The core problem is simple: if chain 1 and chain 2 don't share a settlement layer, you cannot literally move the same token from one to the other. You can only *transform* it. Most bridges do this through a *lock-and-mint* pattern:

- Lock the original asset on chain 1 in a controlled vault.
- Mint a representation (a wrapped token) on chain 2.

The wrapped token is a claim—explicit or implicit—on what's locked in the vault. The bridge is not moving a coin through space. It is creating an IOU on one chain backed by collateral on another.

That's where the risk concentrates: in the vault and the rules that govern it.

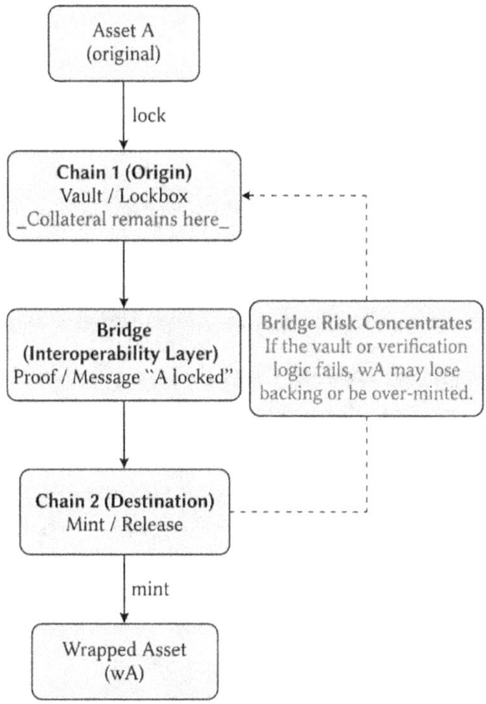

That diagram is the key insight: the bridge is a promise factory. On chain 2, the wrapped token looks like an asset you can trade, lend, and use as collateral. The hard value is still sitting in the vault on chain 1. If the vault is compromised or the bridge's verification logic is fooled, the wrapped token can become a beautifully liquid instrument backed by nothing but disappointment.

Why bridges get hacked: they are high-value and tightly coupled

Bridges are tempting targets for the same reason a bank vault is: they hold concentrated value. But bridges have an additional problem: complexity. A bridge must accurately track events on one chain and reproduce consequences on another. That requires:

- some mechanism for observing chain 1,
- some mechanism for convincing chain 2 that the observation is true,
- and a set of contracts that enforce minting and burning rules.

Each of those mechanisms creates a surface for attack. Some bridges rely on multisig guardians. Some rely on light-client verification. Some rely on a set of validators that sign messages. Each design is a different flavor of trust—and each flavor can curdle if the incentives, implementation, or operational practices are weak.

The harsh truth is that a bridge often becomes a *shadow custodian*. Even when it's marketed as decentralized, it is performing a custody-like function: holding collateral in a vault and issuing claims. If you wouldn't accept a custodian whose controls you don't understand, you shouldn't accept a bridge whose trust model you can't explain in a sentence.

Wrapped assets: useful, but always a little bit second-class

Wrapped assets are an ingenious hack. They allow a token to show up where it otherwise cannot exist, enabling liquidity and composability. They are also, by nature, derivatives: claims on collateral held elsewhere.

This "second-class" status shows up in subtle ways:

- The wrapped token can trade at a discount if the market worries about the bridge.
- Some platforms will not accept wrapped versions as collateral, or will haircut them.
- During stress, redemption can become slow or impossible, even if trading continues.

The last point is the most important. Markets can remain liquid right up until the moment they discover that convertibility is constrained.

That is not unique to crypto; it is a theme of every financial crisis. Bridged assets just provide a crisp, programmable version of it.

Settlement delays in rollups: the ``fast'' chain with the slow exit

Not all cross-domain movement is bridge risk in the classic sense. Layer-2 systems (rollups) can offer fast trading and cheap transactions while ultimately anchoring security to a base chain like Ethereum. That anchoring introduces a new kind of settlement reality: *withdrawal finality*.

Some rollup designs include a challenge period—a waiting window during which withdrawals can be contested if fraud is detected. In practice, this can mean that moving assets from the rollup back to the base chain is slow, even if activity inside the rollup is fast. The operational implication is not theoretical. It affects:

- treasury management (when can you access base-layer liquidity?),
- exchange operations (how do you manage deposits/withdrawals across layers?),
- and risk (what happens to users if they need to exit quickly during stress?).

Speed inside a domain is not the same as speed across domains. Many market mishaps come from confusing the two.

Bridging is not settlement; it is a new obligation layered on top

Settlement finality asks: "Is this transfer irreversible on this ledger?" Bridging asks: "Can this ledger reliably recognize and honor what happened on that other ledger?" The second question is harder because it involves two systems and an interface between them.

That interface is where trust hides. A bridge can have perfectly final settlement on both chains and still fail because the bridge logic minted too many wrapped tokens, accepted a forged message, or lost control of a vault. The danger is that users see "final" on chain 2 and assume that means "safe." Finality on the destination chain only means the mistake is permanent.

Operational heuristics: how professionals think about movement risk

Professionals tend to be boring in the best way. They don't ask, "Is this bridge cool?" They ask questions that sound like a risk committee memo:

- *What is the trust model?* Multisig? Validator set? Light client? Who can upgrade the contracts?
- *Where does the collateral sit?* What exactly is locked, and who can unlock it?
- *What is the failure mode?* Over-minting? Vault drain? Censorship? Stalled withdrawals?
- *What is the exit plan?* If the bridge is paused or attacked, what can users do?
- *How much liquidity do we need on each side?* Movement delays turn into funding needs.

Those questions also illuminate why institutions often prefer to keep core balances on the most secure, most widely supported settlement layer, even if it is slower or more expensive. Cheap movement is not cheap if it introduces a tail risk that can wipe out the savings in one incident.

The human factor: bridges don't just fail technically

Even when the code is solid, bridges can fail operationally:

- governance keys get compromised,
- upgrade processes are rushed,

- monitoring misses anomalies,
- or emergency controls are triggered too late.

In other words, bridge risk is not only about cryptography. It is about *operations under pressure.* That should feel familiar, because it is the same story as custody and venues: the technology creates new capabilities, but the weak point is still the intersection of people, permissions, and money.

A useful mental model: move value like you move inventory

For active traders, moving assets between chains can feel like moving tabs between browser windows. For a treasury team, it should feel like moving inventory between warehouses. Warehouses have locks, insurance, delivery windows, and routing choices. Transfers have confirmation policies, bridge trust assumptions, congestion risk, and sometimes days-long exit timelines.

Once you adopt the inventory mindset, good behavior follows naturally:

- Keep buffer liquidity where you actually need it.
- Don't route critical balances through the newest, least-tested bridge because the fee is lower.
- Treat "wrapped" as a qualifier, not a footnote.
- Assume that in a crisis, bridges will be where liquidity goes to argue with physics.

The headline lesson is simple and slightly uncomfortable: blockchains can settle value with remarkable integrity *within* their own domains, but moving value *between* domains reintroduces trust. Bridges are how that trust is packaged. Some packages are robust. Some are fragile. All of them deserve the same scrutiny you would give any entity that holds your collateral while issuing you a promise.

Chapter 5

Guardrails: Regulation, Compliance, and Risk

If blockchain technology is a high-speed engine driving the future of finance, regulation is the braking system that keeps the car from flying off a cliff. In this chapter, we step away from the hype to look at the sober reality of the 'Wild West' transitioning into a civilized economy. We explore why regulators care so much about your digital tokens, why the taxman is watching your wallet address closer than you think, and how to spot a scam before it drains your account. This isn't just about following rules; it's about understanding the safety mechanisms that make digital assets usable for the rest of us.

5.1 The Regulatory Perimeter: Is It a Stock, a Commodity, or Just Magic Internet Money?

A digital token can look like a baseball card at breakfast, a stock certificate at lunch, and a casino chip by dinner. That shapeshifting quality isn't just annoying for lawyers; it's the entire regulatory problem. Finance law is built around buckets—*security, commodity, currency, deposit, derivative*—and each bucket comes with a different set of promises to the public. Different disclosures. Different gatekeepers. Different cops. Digital assets show up wearing a hoodie that says, "I'm none of the above," while quietly doing the same economic work those buckets were invented to contain.

Regulators don't actually care what you *call* the thing. They care what the thing *does*. That idea sounds obvious until you realize how much of modern crypto marketing is basically a thesaurus. "Not a stock—a token." "Not interest—a reward." "Not dividends—yield." "Not a company—a community." The legal system has a blunt response to this: *economic reality over labels*. If it walks like an investment and quacks like an investment, it's going to be treated like one, even if the whitepaper insists it's a "utility."

The U.S. in particular runs a kind of regulatory *federalism of finance*. There isn't one master agency that wakes up, stretches, and says, "Today I regulate all digital assets." Instead, several agencies have jurisdiction over different slices of behavior. If your token sale looks like capital formation, securities law gets interested. If your product offers leveraged trading to regular people, commodities law gets interested. If you're moving value for customers, financial-crime rules get interested. If you're facilitating payments, state money transmission regulators get interested. The same token can trigger multiple tripwires depending on how it's sold, traded, and used.

The result is a perimeter problem: where exactly does "innovation" end and "regulated finance" begin? Crypto fans sometimes talk about regulation as if it's a tax paid to the gods of bureaucracy. The more accurate view is that regulation is the price of entry to the most valuable thing in finance: *other people's trust at scale*. Investor protection is not a vibe; it's a set of mechanisms that make it harder for smart people with bad intentions to vacuum money out of crowds.

Securities law is the sharpest tool in that kit. A security isn't just a share of stock in the plain, familiar sense. U.S. law treats many arrangements as securities if they are *investment contracts*—a phrase that sounds like paperwork but functions like a net. The famous test (born from a citrus-grove case that refuses to die) asks whether people are putting in value, joining a common enterprise, expecting profits, and relying primarily on someone else's efforts to make those profits happen. If that describes your token distribution, the law tends to treat it less like a collectible and more like raising money from the public.

That reliance piece is the tell. If your token's value depends on a founding team shipping features, negotiating exchange listings, buying back tokens, managing treasuries, or "driving adoption," you've basically described management. Management implies an audience. An audience implies asymmetric information. Asymmetric information is where fraud breeds. Securities regulation exists to force sunlight into that gap: mandated disclosures, liability for false statements, restrictions on who can sell what to whom, and rules for intermediaries who touch the transaction. None of that is very exciting—until you've watched people lose their savings to a token that was, functionally, a startup equity round dressed up as a game.

Here's the uncomfortable bit for crypto idealists: "decentralized" is not a spell that breaks the law. A project can have a community and still have a small group pulling the levers that matter. Sometimes the token distribution is decentralized in the way confetti is decentralized: it's everywhere, but someone still threw it. Regulators look for the "active participant"—the person or group whose ongoing efforts are the reason buyers think number-go-up. If that participant exists, it's hard to argue purchasers are simply buying a consumable good. They're buying a bet on execution.

Even when a network becomes more genuinely decentralized over time, the early phase can still look like a classic capital raise. That temporal nuance matters. Many digital assets begin life as something very close to a startup financing mechanism: money comes in, tokens go out, and the team uses the proceeds to build. Later, the token may circulate more like an open commodity used for network fees, collateral, or governance. That "evolution" is a real phenomenon, but it doesn't retroactively rewrite history. Investor-protection rules are often most needed at the beginning—when hype is highest, information is scarcest, and the line between ambition and fiction is thinnest.

The SEC's interest in token offerings isn't a philosophical dispute about technology; it's a practical question about public fundraising. If you are raising money from the public with promises (explicit or implied) that your work will create profit for them, securities law wants to turn the lights on. That means registration or an exemption, disclosure documents that can be sued over, and limits on how you can market the thing. To crypto entrepreneurs, that can feel like trying to paint a mural while wearing boxing gloves. To the public, it's one of the few systems designed to prevent the same old playbook: charismatic promoter, complicated product, missing disclosures, sudden collapse.

Commodities law enters the story from a different angle. A commodity is not just corn and oil; it's a broad concept that includes goods and articles traded in markets, and in U.S. practice it often captures certain virtual currencies that function more like raw financial material than a claim on a business. Commodity regulation tends to focus less on the issuer and more on the integrity of the marketplace: fraud, manipulation, and the plumbing of derivatives. If something trades like a commodity—especially if there's futures or leveraged trading attached—the CFTC starts paying attention.

This creates one of the most confusing optics for newcomers: how can something be a security *and* a commodity? Sometimes the answer is that different parts of the ecosystem are being regulated, not that the asset has a single eternal identity. A token sold in an initial offering can be treated as a securities transaction because it looks like fundraising. That same token, once broadly distributed and traded, can behave like a commodity in secondary markets, especially when it becomes the underlying reference for futures, swaps, or margined products. The *transaction* and the *market structure* matter at least as much as the object itself.

This is where the "regulatory perimeter" starts to feel less like a line and more like a weather system. The perimeter shifts depending on who is interacting, what they're promised, and what leverage or intermediaries are involved. A spot purchase of a token for immediate delivery may trigger fewer commodity-law obligations than a leveraged "trade" that looks suspiciously like a derivative sold to retail customers. The same app can move in and out of different regimes simply by changing settlement timing, custody arrangements, or whether it offers margin. In traditional finance, those design choices are boring. In crypto, they're product features. Regulators see them as legal facts.

Then there's the category that crypto loves to invoke: *currency*. People say "digital currency" the way they say "digital camera," implying it's just the modern version of an old thing. But legally, "currency" is not a compliment you can award yourself. In the U.S., "legal tender" is a specific status tied to sovereign authority. Most crypto assets are not that. Calling something a currency may describe how fans *want* it to behave, but it doesn't automatically change which laws apply when it's marketed, sold, or intermediated.

Stablecoins complicate this further because they are designed to feel like money. A well-run stablecoin can behave like a payment instrument; a badly run one can behave like a fragile money-market fund with none of the guardrails. From a regulatory point of view, stablecoins are a crossroads: payments, banking, securities-like features (if yield or profit expectations are baked in), consumer protection, reserve transparency, and operational resilience. The classification question isn't academic. If the public treats a stablecoin as cash and it breaks, the damage looks less like a failed startup and more like a run on a shadow bank.

NFTs add their own brand of confusion, largely because they carry an *aesthetic* that distracts from their economics. A single NFT that is genuinely a one-off collectible can resemble art or memorabilia. But a large NFT project marketed with roadmaps, token-gated perks, floor-price obsession, and a core team "building value" starts to rhyme with an investment scheme. Fractionalization, revenue-sharing, and pooled purchasing can push the structure even closer to something securities law recognizes. The picture may be of a cartoon ape; the transaction can still be an investment contract.

Tokenization of traditional assets is where the perimeter debate gets even more revealing. If you tokenize equity in a company, you haven't escaped securities regulation; you've just changed the format. A tokenized share is still a share. The interesting question becomes operational: can the market infrastructure (custody, transfer restrictions, corporate actions, record ownership) meet the same investor-protection expectations with new rails? In other words, tokenization isn't a loophole—it's a proposal to modernize the machinery while keeping the safety standards.

The perimeter also depends on *who* is doing the selling. The same asset can be sold in ways that feel radically different in terms of investor protection. A direct sale by an issuer with glossy promises is one thing. A secondary trade between two users is another. A centralized exchange listing is yet another: the exchange becomes a powerful intermediary that can either filter bad assets or amplify them. In traditional markets, intermediaries like broker-dealers and exchanges are heavily regulated precisely because they are distribution engines. Crypto platforms, at least historically, have often wanted the power of distribution without the responsibilities attached to that power.

That responsibility isn't just about paperwork. It's about conflicts of interest, surveillance, custody, and market integrity. If a platform both lists a token and trades against customers, that's not "innovation"; it's a conflict that traditional finance learned to fear the hard way. If a platform can halt withdrawals because it mismanaged customer assets, the problem isn't that the technology is new. The problem is that the oldest rule in finance—*don't lose the money*—still applies.

Investor protection, at its best, is boring. It's a set of design constraints that make disasters less frequent and less catastrophic. Securities registration forces issuers to make standardized disclosures and gives investors legal remedies when those disclosures are false. Broker-dealer rules impose suitability and best-execution expectations in many contexts. Exchange regulation aims to prevent wash trading, manipulation, and unfair access. Custody rules are meant to separate customer assets from a firm's balance sheet so that "your money" is not a synonym for "their working capital." These mechanisms are the guardrails people only notice when the car is already in the ditch.

Crypto's counterargument is that some of these protections are ill-fitted to open networks. That critique isn't entirely wrong. A decentralized protocol doesn't have a CEO who can sign an S-1 registration statement. Code is global; rules are national. Token holders can be everywhere and nowhere. But the uncomfortable reality is that many "decentralized" projects still have human choke points: foundations, core developers, treasuries, front-end websites, governance delegates, and market makers. Investor-protection law gravitates toward those choke points because that's where accountability can attach.

This is why the classification debate can't be reduced to a meme about "regulators don't understand the tech." Often, they understand the tech just fine. What they're doing is pattern matching the economic incentives. Who benefits if buyers show up? Who controls the messaging? Who holds the treasury? Who can change parameters? Who has information before everyone else? Once you ask those questions, a token starts to look less like "magic internet money" and more like a familiar arrangement wearing unfamiliar clothes.

The "duck test" framing works because it cuts through the word games. If it trades like a stock—because people buy it primarily to profit from price appreciation—and it funds a venture like a stock—because proceeds finance a team that builds value—then pretending it is merely a "utility token" is like selling season tickets to a stadium that doesn't exist yet and insisting you're not raising money, you're selling *future enjoyment*. Maybe you are. But if the stadium never gets built, the law wants someone answerable.

None of this means every token is a security. Some tokens really do function more like commodities or payment instruments. Some networks reach a level of dispersion where there is no meaningful managerial team to rely on. Some assets are closer to digital collectibles than investments. The point is that classification is not a vibe-check; it's a facts-and-circumstances analysis. And facts are slippery in a market where a single tweet can turn "community experiment" into "number-go-up opportunity" in five minutes.

A useful mental model is to think of regulation as a map of *risks*, not just asset types. Securities law is the map for capital-raising risk and information asymmetry. Commodities law is the map for market-manipulation risk, especially around derivatives and leverage. Payments and banking regulation is the map for run risk and consumer protection. Financial-crime rules are the map for illicit finance risk. Tax rules are the map for, well, the government's share. Digital assets don't neatly live in one region of the map; they cross borders constantly.

That border-crossing is also why "regulatory arbitrage" has been such a persistent crypto strategy. If one country draws a perimeter you don't like, you incorporate somewhere else, route users through a different entity, block IP addresses, and hope the internet's natural friction keeps regulators away. Sometimes it works for a while. But if your customers, marketing, founders, or trading activity are meaningfully connected to a jurisdiction, the perimeter has a way of finding you. The market may be global; enforcement tends to be local and very attached to where people live, bank, and keep servers.

For investors, the classification question isn't trivia—it changes what protections you can realistically expect. When an asset is treated as a security, the issuer may be required to provide disclosures and may face liability for misleading statements. When a platform is regulated as a securities intermediary, there are expectations around custody, supervision, and conflicts. When something falls into a gap, you may be operating in a zone where your primary protection is your own skepticism and operational discipline. That doesn't mean "unregulated" equals "bad," but it does mean "buyer beware" is doing more work than people admit.

It also changes how you should interpret the marketing. If a token is pitched with language that sounds like venture capital—"early," "ground floor," "upside," "team," "partnerships," "roadmap," "token burns," "price support," "treasury management," "institutional interest"—you're not being sold a product like a hammer. You're being sold a story about future value. Stories can be true. Stories can also be cheap to manufacture. Securities regulation exists largely because humans are surprisingly easy to persuade when a chart points upward.

The deeper lesson is that regulation isn't anti-innovation; it's anti-*asymmetric* innovation—the kind where complexity protects the seller more than the buyer. Crypto can absolutely modernize finance. But modernizing finance without modernizing investor protection is like building a faster car and removing the seatbelts because they're "legacy." The perimeter debate is really a debate about where the seatbelts should attach when the vehicle's shape keeps changing.

So when someone asks whether a token is a stock, a commodity, or magic internet money, the most honest answer is: it depends on the economic deal being offered and the structure around it. If you're being asked to fund a team and trust their efforts, securities logic is in the room. If you're trading something that behaves like a broadly used raw asset, commodities logic is in the room. If you're using it like money, payments and banking logic is in the room. And if you're hearing that it's *none* of these because it has a clever new name, you're not watching innovation. You're watching a sales technique.

5.2 Compliance and Financial Crime: Why You Can't Stay Anonymous Forever

Crypto's original sales pitch had a romantic villain: the gatekeeper. Banks asked nosy questions. Payment networks could say no. Governments could freeze accounts. The fix, according to the early mythology, was math. Replace permission with proof-of-work, replace account managers with private keys, and money becomes something you *hold* rather than something you *ask to use*. That story still has emotional power—right up until the moment the real economy shows up, wearing a name badge that reads *Compliance.*

Compliance isn't a vibe-killer; it's the operating system that lets strangers move large amounts of value without constantly assuming the other side is a criminal. Anti-Money Laundering (AML) rules and Know Your Customer (KYC) checks exist for an unglamorous reason: financial rails are the favorite highways for drug trafficking, ransomware, sanctions evasion, corruption proceeds, and fraud. When a system becomes useful for normal people, it also becomes useful for determined criminals. That's not cynicism—it's just the economics of convenience.

The awkward truth is that most people don't interact with blockchains directly. They interact through choke points: exchanges, stablecoin issuers, payment processors, hosted-wallet apps, on-ramps, off-ramps. These are the places where crypto touches bank accounts, cards, and payroll. They are also the places where regulators can credibly demand, "Who are you, and where did that money come from?" If you thought crypto would eliminate gatekeepers, what it often did was *privatize* them. The velvet rope didn't vanish; it changed owners.

KYC is the part you feel. It's the selfie, the passport scan, the "why are you sending funds?" questionnaire that arrives exactly when you're most excited to buy your first token. KYC's job is identity: linking a real-world person (or business) to an account, so that the system can enforce sanctions lists, investigate suspicious behavior, and—in the bluntest terms—make it possible to hold someone responsible. AML is the broader program behind it: policies, monitoring, reporting, training, audits, and controls designed to reduce financial crime. KYC is the front door; AML is everything else that keeps the house from becoming a safehouse.

It's worth being precise about what "anonymous" means in crypto, because the word gets abused. Most major blockchains are not anonymous; they are *pseudonymous*. A wallet address is a string of characters, not a name—but the transactions are public, permanent, and linkable. The minute your address touches a regulated intermediary, a payroll provider, a bank transfer, a centralized exchange account, or even a merchant that keeps records, pseudonymity starts leaking like air from a punctured tire. One clean KYC event can illuminate a huge web of historical activity.

This is where crypto's most counterintuitive feature becomes compliance's best friend: the ledger doesn't forget. In the traditional banking system, a lot of data lives in separate private databases. Investigations often involve subpoenas, reconciliation, and waiting for someone to cooperate. On-chain, the transaction graph is sitting there, already normalized, already time-stamped, already showing flows. Analytics firms don't need to "break the blockchain" to do their work. They just need to connect addresses to entities and then follow the money the way you'd follow footprints in fresh snow.

So why does compliance feel harsher in crypto than in banking? Because crypto businesses often meet the real compliance world *after* growing quickly. In banking, AML controls are baked into the institution from day one. In crypto, a startup can go from side project to multi-billion-dollar throughput before it has mature governance, risk functions, or even clarity on which regulator is in charge. When regulators arrive, the response can be abrupt: registrations, enforcement actions, and a sudden shift from "move fast" to "document everything."

In the United States, a lot of the AML perimeter runs through the Bank Secrecy Act (BSA) framework and its implementing rules. The key practical concept is that many crypto intermediaries can be treated as money services businesses (MSBs), including money transmitters, depending on what they do. The law cares less about whether your app calls itself "DeFi" and more about whether you are in the business of transmitting value for others, exchanging one form of value for another, or administering a convertible virtual currency. When you're inside that perimeter, you're signing up for a full-time job called "not being a conduit for crime."

That job has a rhythm. A compliant firm runs customer due diligence (CDD): collecting identifying information, verifying it, and understanding the customer relationship. It screens customers against sanctions and watchlists. It monitors transactions for suspicious patterns. It keeps records for prescribed periods. And when it sees activity that looks like money laundering, fraud, or sanctions risk, it files reports (often Suspicious Activity Reports, or SARs) with financial intelligence units. None of this is theater. Filing SARs is one of the ways the system produces investigatory leads without requiring every business to be a law-enforcement agency.

The phrase "risk-based" gets thrown around so often it can sound like a corporate screensaver, but it matters. AML is not supposed to be a single checklist applied identically to everyone. A small, local business buying $500 of bitcoin and withdrawing to a personal wallet is not the same risk profile as a high-volume account funneling funds through mixers and rapidly cycling assets through multiple chains. A good compliance program assigns different levels of scrutiny to different behaviors. A bad program treats KYC like a one-time password reset: painful, performative, and quickly forgotten.

Sanctions compliance is its own cold-blooded layer. You can follow every AML rule in the world and still end up in trouble if you move value for a sanctioned person, group, or jurisdiction. Sanctions are not about the *source* of funds being criminal proceeds; they're about national-security policy. And unlike some financial-crime concepts that rely on intent, sanctions compliance often functions like a strict-liability world: if you did the prohibited thing, the problem exists whether or not you "meant to." That's why serious crypto firms invest heavily in screening, blocking, and escalation procedures. The chain doesn't care about geopolitics. Regulators do.

The compliance friction is also the price tag for institutional participation. Big asset managers, public companies, banks, and payment networks have reputations that are harder to replace than code. They cannot afford to plug into a system where the standard defense against financial crime is "don't worry, we're decentralized." If you want pension funds to treat tokenized assets as normal instruments, you need controls that resemble normal finance: onboarding standards, transaction monitoring, audit trails, and a clear ability to respond when something goes wrong. The market's most conservative players are not being stubborn; they're being rational about career risk.

Crypto's objection is predictable: KYC breaks privacy and undermines the open-access promise. That complaint deserves more nuance than it usually gets. There is a real tension between privacy and surveillance, and the last decade has shown that data collected "for safety" has a way of being misused, breached, or repurposed. When an exchange holds millions of identity documents, it becomes a honeypot for hackers. Compliance can reduce one category of risk while increasing another. The correct response isn't to abandon KYC; it's to treat customer data as a high-value liability and build controls that match the sensitivity: minimization, encryption, access controls, vendor management, and incident response.

There's also a philosophical mismatch: blockchains are designed for *trustless* verification, but compliance is built around *trust plus accountability*. AML doesn't just ask, "Did the transaction happen?" It asks, "Should this transaction have happened, and can we explain it?" The first question is technical. The second is social and legal.

You can't answer it with cryptography alone, because the meaning of "should" depends on laws, policies, and evolving geopolitical realities.

That mismatch shows up in the DeFi conversation. DeFi applications often argue they are merely code: a set of smart contracts on a blockchain that anyone can use. But code has authors. Protocols have governance. Interfaces have operators. Liquidity incentives have designers. When profits exist, people manage the levers that create them. Regulators look for the humans behind the "autonomous" system because financial-crime frameworks are fundamentally built around responsible parties. The idea that a protocol can operate at scale, touch the real economy, and be accountable to no one is not a neutral design choice; it's a bet against the entire arc of financial regulation.

Even if you never touch a regulated exchange, "staying anonymous forever" runs into a simpler problem: you eventually want to *use* your money. That means buying something, paying someone, or converting back into fiat to cover rent and taxes. Each of those moments creates linkage opportunities: merchant records, bank transfers, delivery addresses, IP logs, device fingerprints, and counterparties that are themselves regulated. A wallet address can be nameless, but the world is full of receipts.

The newest reality check is that compliance is getting industrialized. Early crypto felt like cash in a trench coat; today it increasingly resembles brokerage infrastructure, complete with information reporting and standardized statements. Even without discussing tax details, the direction is clear: regulators want the ecosystem to generate structured data about transactions, the same way traditional finance does. When that happens, anonymity doesn't disappear because the blockchain was "cracked." It disappears because the financial system wraps the blockchain with reporting obligations and audit trails.

For consumers, the practical implication is subtle but important: the safest way to interact with digital assets often involves intermediaries, and intermediaries are regulated. That means your experience will feel less like the anarchic internet and more like opening a financial account. If a platform is willing to take your money without asking who you are, that can be a feature—but it can also be a warning label. "No questions asked" is not a compliance innovation; it's the marketing slogan of every laundering channel ever invented.

For businesses, compliance becomes a competitive advantage once the market matures. In the early phase, firms sometimes treat compliance as a tax on growth and try to do the minimum. In the later phase, the firms that can demonstrate credible controls get access to better banking relationships, more reliable liquidity, and institutional clients who actually pay their bills. The irony is delicious: the best way to scale an industry that wanted to eliminate trust is to become trustworthy in the most traditional way possible.

A good compliance culture also changes how products are designed. When you expect to monitor transactions and respond to risk, you start caring about traceability, governance, and control points. You care about whether a protocol can freeze stolen funds (a controversial feature, but a real one in some designs). You care about whether upgrades are transparent and auditable. You care about how quickly you can respond to hacks, scams, and sanctions updates. These are not mere legal concerns; they are engineering requirements that shape what "mass adoption" actually looks like.

None of this means the future is one where every wallet is tattooed with a Social Security number. There are plausible paths toward privacy-preserving compliance: cryptographic proofs that someone has passed checks without revealing every detail, selective disclosure, and better compartmentalization of identity data. But those are hard problems. And until they are solved at scale, the default approach will remain the one regulators know how to supervise:

identify customers at the edges, monitor flows, report suspicious activity, and enforce sanctions.

So the honest answer to "Why can't I stay anonymous?" is: because anonymity at scale is indistinguishable from a financial crime feature. You can build technology that resists control. You can also build a market that invites real savings, real salaries, real corporate treasuries, and real public trust. Doing both at the same time is like trying to run a city where nobody has an address and everyone expects the fire department to find them anyway.

Compliance is the part of crypto that admits it is growing up. The party doesn't end; it gets a bouncer, a guest list, cameras at the door, and a very boring manager who knows exactly where the exits are. That manager is not there to ruin the night. They're there because, sooner or later, someone tries to start a fire.

5.3 The Taxman Cometh: The Nightmare of Record Keeping

The blockchain might be a marvel of cryptography, but the IRS is impressed by a different feature: it keeps receipts forever. Crypto culture once flirted with the idea that a public address was a kind of invisibility cloak. Tax law looks at that and shrugs. In the U.S., most digital assets are treated as *property* for federal income tax purposes, which is the same bucket that catches stocks, real estate, and a surprisingly large portion of adult headaches. Property rules are relentless: when you dispose of property, you generally measure gain or loss. And "dispose" is a bigger verb than most people expect.

A lot of taxpayers learn this the hard way because crypto transactions *feel* like spending money. You tap "Pay," the merchant gets something valuable, and you walk away with coffee. But if you paid with Bitcoin (or most other tokens), you didn't just buy a latte. You *sold* property—even if you never touched dollars—and then used the proceeds (in economic terms) to buy the latte. That mental flip is the whole nightmare: in fiat, everyday life is financially invisible. In crypto, everyday life can become a spreadsheet.

Taxable events are the moments the IRS cares about because they're the moments where economic value changes hands in a way that can produce gain or loss. Selling a token for dollars is the obvious one. Exchanging one token for another counts too, because you disposed of the first asset and acquired a new one. Paying for goods and services with a token is another. Receiving tokens as compensation is another. Staking rewards, airdrops, certain yield arrangements—the details can get technical fast, and the right answer depends on facts and timing—but the big theme is simple: the tax system has an allergy to unrecorded value transfer.

The problem isn't that these rules are conceptually difficult. The problem is that crypto turns normal consumer behavior into a high-frequency trading diary. If you bought BTC in 2021, swapped some for ETH in 2022, bridged assets in 2023, hopped through a DeFi protocol in 2024, then spent a little in 2026, you now have a timeline that looks less like "personal finance" and more like an investigative whiteboard. Each hop potentially creates a new cost basis, a new acquisition date, and a new disposition that needs to be priced in dollars at the time it happened.

Cost basis is the number that quietly controls the entire story. Basis is, roughly, what you paid to acquire an asset (including certain fees), and it determines whether you have a gain or a loss when you dispose of it. If you bought 0.05 BTC for $1,500 and later spend it when that slice is worth $2,000, that $500 difference is a capital gain before you even think about the coffee. If you don't know your basis, you don't know whether you owe taxes, and you don't know whether your "profit" is real or just a mirage created by missing data.

Holding period adds another layer of personality. The U.S. tax code has different rates for short-term versus long-term capital gains, and the dividing line is time: hold for more than a year and you often get a better rate; hold for a year or less and gains are generally taxed like ordinary income. In crypto, where people can transact multiple times in a day, the holding period is easy to lose track of and expensive to ignore. A token sold 366 days after purchase can be meaningfully different from the same token sold 364 days after purchase. Two days can be money.

The practical shock comes when people realize that "I only used crypto a little" can still mean "I created dozens of taxable events." A "little" might be five swaps, two stablecoin conversions, three NFT mints, a sale, and a payment to a friend. The blockchain makes those actions frictionless; the tax system makes them *count*. If you have ever wondered why traditional finance is full of intermediaries, paperwork, and batches, here's one answer: the system evolved to record activity in ways that match reporting. Crypto evolved to execute activity. Reconciling those two philosophies is what produces the April migraine.

Immediately after you understand taxable events, it becomes clear why a cash purchase feels like a warm blanket and a crypto purchase feels like homework.

Buying Coffee: Tax Simplicity vs. Crypto Paper Trail

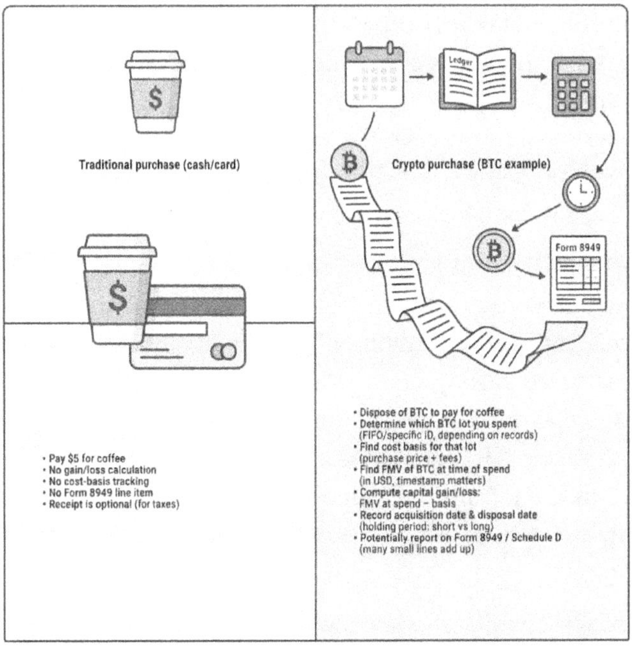

Key idea: using BTC can be a spending event and a taxable disposition at the same time.

The table makes the point, but it doesn't show the cruelest detail: it's not just *one* calculation. It's one calculation per lot. Crypto users often acquire the same token at many different prices over time, which means "your Bitcoin" is really a pile of separate purchase lots with separate bases and acquisition dates. When you later dispose of some amount, you need a method to decide which lot you used. If you can specifically identify units (and have records that would satisfy scrutiny), you may be able to pick lots strategically. If you can't, you're often stuck with default conventions that may produce a larger tax bill than you expected.

Now add decentralized finance, where your assets don't sit quietly in a wallet like gold in a sock drawer. They get wrapped, staked, deposited, lent, borrowed against, pooled into liquidity positions, or bridged between chains. Each action has a business purpose inside the protocol, but tax reporting wants a simpler story: what did you give up, what did you receive, and what was each worth in dollars at that time? DeFi is a machine for generating complex answers to those questions. Even when the legal interpretation of a particular DeFi action is contested or still developing, the recordkeeping burden arrives immediately, because you still need the underlying data.

NFTs bring a different kind of complexity: pricing. With a publicly traded token, you can usually point to a reasonably objective market price at a timestamp. With NFTs, valuation can be messy. Floor prices can move quickly, liquidity can vanish, and the "value" of an NFT may depend on traits that don't have an obvious market quote. If you swap one NFT for another, or pay an NFT as compensation, you still need a fair market value in dollars to compute income or gain. The tax system asks for a number even when the market is shrugging.

Stablecoins are where many people expect relief: "If it's pegged to a dollar, surely taxes get easier." They can get *less* dramatic, but not necessarily easier. If you're constantly swapping stablecoins, the gains may be small, but the number of transactions can be large. A thousand micro-gains are still a thousand lines of data, and the human brain does not naturally enjoy tracking pennies across a year's worth of on-chain activity. The spreadsheet doesn't care that each item is tiny; it cares that each item exists.

As of March 2, 2026, the U.S. reporting environment has also started to harden in a way that changes the texture of the problem. Broker reporting for digital assets has been phasing in: beginning with transactions on or after January 1, 2025, certain brokers are required to report gross proceeds on a new form, Form 1099-DA. Basis reporting is phased in for certain covered transactions on or after January 1, 2026. This isn't a philosophical statement; it's a practical one: more third-party statements mean less room for "I forgot" and more automated matching. The tax system is increasingly designed so that your return is not the only record of your activity.

That shift cuts both ways. For taxpayers who mainly use custodial platforms, broker statements can reduce the chaos. You'll get paperwork that resembles the traditional brokerage world: summarized proceeds and, increasingly, cost basis information for covered assets. But there's a catch that surprises people: crypto activity is often fragmented across platforms and wallets. One exchange knows what happened inside its walls. Your self-custody wallet knows what happened on-chain, but it doesn't produce a tax statement. Your DeFi protocol knows what happened in its contracts, but it won't mail you a neat form. So you end up stitching together a quilt made of partial truths.

Even within the broker reporting world, "covered" versus "noncovered" is where the fine print lives. A broker may have to report proceeds even when it cannot reliably report basis, especially for older acquisitions, transferred-in assets, or lots acquired before basis reporting applies. That means you can receive a statement that feels authoritative but still leaves the most important number—basis—as your responsibility. People see a 1099-like form and assume the hard part is done. Sometimes it isn't. The form can be a map with missing streets.

That reality makes recordkeeping less like "save your trade confirmations" and more like building a personal ledger that can survive platform changes. Exchanges merge. Apps shut down. Wallet providers disappear. APIs change. If your only record of a 2021 purchase is an email you can't find and an app that no longer exists, you haven't just lost nostalgia; you've lost basis. And losing basis is rarely convenient. When you can't substantiate basis, you can end up paying tax as if your basis were zero—the tax equivalent of being charged twice for the same meal because you threw away the receipt.

The psychological trap is that crypto users often treat transactions as reversible in meaning even if they are irreversible on-chain. "I was just moving funds around." "I was just swapping to access a pool." "I was just bridging." The tax system doesn't grade your intention; it grades your disposals and acquisitions. That mismatch between intent and treatment is why sophisticated market participants invest in systems that capture data at the moment of execution: timestamps, transaction IDs, wallet addresses, token quantities, fees, and USD values. Waiting until February to reconstruct a year's worth of activity is like trying to rebuild a demolished building from a handful of bricks and your memory of where the kitchen used to be.

The good news is that the nightmare has patterns, and patterns can be managed. The most powerful habit is boring: pick a tracking method early and stick to it. Whether you use dedicated software, a spreadsheet, or professional help, the goal is the same—create a single source of truth that reconciles custodial statements with on-chain reality. Keep exchange CSVs when you download them. Keep wallet addresses labeled. Record which addresses are yours. Save the transaction hashes for large or unusual events. Make notes in plain English when you do something complicated, because "0x9f..." will not jog your memory in ten months.

There's also a strategic insight that feels almost impolite to say out loud: if you want tax simplicity, you may have to sacrifice a little technological purity. Spending directly from a volatile token creates a trail of tiny taxable events; spending from dollars is boring, and boring is easy. Using a custodial platform may produce cleaner statements than an ecosystem of fresh wallets and experimental protocols. Some people will gladly accept the paperwork as the price of freedom. Most people, when faced with a shoebox of transaction IDs, discover they prefer freedom *and* sleep.

The punchline is that the tax system doesn't care whether the blockchain is futuristic. It cares whether you had an accession to wealth, whether you disposed of property at a gain, and whether you can substantiate your numbers. Crypto can be programmable money, decentralized infrastructure, and a new asset class all at once.

For taxes, it is often something more mundane: a machine that turns your financial life into data. The taxman doesn't need to chase you through alleyways. He just needs you to forget where you put the receipts.

5.4 Operational Risk and Fraud: Don't Get Rugged

The cruel joke of self-custody is that it turns everyone into a small financial institution without handing out the corresponding staff. In traditional finance, "operational risk" is what happens when back offices mis-key a trade, a custodian misroutes a wire, or an employee clicks the wrong attachment. In crypto, the back office is you. The custodian is you. The employee with the itchy click-finger is also you. That doesn't mean crypto is uniquely dangerous; it means the safety features people take for granted in banking are no longer automatic. They're optional add-ons, and sometimes they're missing.

Fraud in crypto also has a special accelerant: finality. A credit card charge can be disputed. A wire might be recalled if everyone moves fast and the other bank is cooperative. A blockchain transfer is more like dropping cash into a storm drain. You can watch it flow away in real time on a block explorer, which is a kind of modern tragedy: perfect visibility, minimal reversibility. That single design feature changes the entire scam economy. Criminals love payment methods that can't be clawed back.

The other accelerant is that crypto collapses distance between "investment" and "payment." Scammers don't just persuade you to buy a bad asset; they persuade you to *send* value to them. Once they get you to send crypto, there is no billing department to call. That's why so many modern scams are built less like financial products and more like psychological funnels: get attention, manufacture trust, create urgency, then extract a transfer.

The most important self-defense move is to stop thinking of "risk" as a single blob. Crypto risk comes in different flavors that require different countermeasures: platform risk (can the exchange freeze or fail?), custody risk (can you lose keys?), transaction risk (can you be tricked into signing something?), and smart contract risk (does the code do what you think it does?). A person who keeps coins on a reputable exchange has concentrated risk in the platform. A person who self-custodies has concentrated risk in their own operational habits. DeFi users add smart contract risk on top, like a second floor built quickly.

Start with the scams that don't need any code at all. The most profitable attacks are often pure social engineering: phishing emails, fake support agents, impersonated founders, "account verification" pages that harvest seed phrases. Crypto's user experience creates a perfect stage for this because legitimate systems genuinely *do* require you to do strange things: copy addresses, confirm networks, manage approvals, sign messages. Scammers don't need to invent weird behavior; they just need to insert themselves into it.

The seed phrase is the crown jewel. A seed phrase (or recovery phrase) is the master key that can regenerate your wallet. Anyone who has it can usually take everything. Real services do not need your seed phrase. Not exchanges. Not wallet support. Not "blockchain validation." Not the polite person in your DMs offering to help. Asking for it is like asking for the deed to your house "just to verify ownership." The only correct response is to close the conversation.

Phishing has also evolved beyond the obvious misspelled email. Scammers buy ads that appear above real search results. They clone websites so precisely that even cautious users can be fooled. They compromise social media accounts and post "official" links. They spin up Telegram and WhatsApp groups with thousands of bots to simulate legitimacy. If it feels like a casino built inside a theater set, that's because it is: the scammers are manufacturing an environment where your brain relaxes, and relaxation is when people sign.

The fastest way to regain control is to build "friction rituals" into your routine. Crypto is thrilling because it's fast; safety often means deliberately slowing down. Type URLs manually or use bookmarks you created yourself. Treat inbound links as hostile, even when they come from someone you know-because "someone you know" might be an account takeover. Never install random browser extensions to "fix" wallet issues. If "support" contacts you first, assume it's a scam; real support waits for you to contact them through an official channel.

Then there's the category of fraud that *uses* the market itself as the weapon: rug pulls. A rug pull is the crypto-native version of an exit scam, and it comes with a marketing soundtrack. The project launches with a token and a story: community, utility, memes, philanthropy, "revolutionary" tech. Liquidity appears on a decentralized exchange, the price starts to move, influencers notice, and suddenly the token has a chart-because nothing makes a story feel true like a chart.

Rug pulls aren't all the same, and that matters because the warning signs differ. Some are 'transaction' rugs: the code is not necessarily malicious, but insiders control the liquidity or hold most of the supply and dump into buyers. The asset is less like a fair market and more like a carnival game where the operator decides when the prizes disappear. Other rugs are 'contract' rugs: the smart contract includes features that let developers change key parameters, block selling, mint new tokens, drain funds, or otherwise rewrite the rules after you've bought in. The first type is about market structure. The second is about code risk. Both end with the same sound: silence in the Discord.

The anatomy is depressingly repeatable. A token is created and paired with a popular asset to form a liquidity pool. Early buyers push up the price. The rising price becomes marketing content. More buyers arrive, usually retail, because the whole point is to recruit liquidity that didn't exist before. At some point, the insiders remove liquidity or dump a concentrated position, and the price collapses because there was never real depth-just a shallow pool lit with neon. People don't lose money because they can't understand liquidity math; they lose money because the social atmosphere convinces them liquidity is a permanent feature rather than a temporary setup.

A blunt but useful question for any new token is: *Who can change the rules after I buy?* In DeFi, "the rules" are often encoded in admin keys, upgrade mechanisms, and governance controls. An upgradeable contract can be a legitimate way to patch bugs; it can also be a loaded gun pointed at users. If a protocol can be upgraded by a small group, you are trusting that group. Trust can be a rational choice. It is not the same thing as "trustless finance," no matter how many times the website says it.

Smart contract risk is where sophisticated people make surprisingly human mistakes. The code might be public, but "public" is not the same as "understood." Even audited code can fail-audits reduce risk; they do not delete it. Bugs can be subtle: rounding errors, price oracle manipulation, flawed access control, reentrancy vulnerabilities, economic exploits that are not strictly "bugs" but still drain value. DeFi is a place where finance meets adversarial game theory. If there is a profitable edge case, someone will eventually find it.

The most dangerous feeling in DeFi is "I've used it a few times and it worked." That's not proof of safety; it's proof you haven't been the target yet. Many exploits happen when market conditions change, liquidity shifts, or an attacker assembles a set of transactions that normal users would never try. Systems can run smoothly for months and then fail in one block. Operationally, that means you should size positions assuming that failure is not only possible but eventually likely somewhere in the ecosystem.

Approvals are another quiet hazard. Many tokens require you to "approve" a smart contract to spend them on your behalf. That's normal in ERC-20 style systems, but it creates a permission layer users rarely revisit. An unlimited approval is like giving a valet a copy of your house key that never expires. If the contract (or an upgrade, or a compromised admin key, or a malicious front end) turns hostile later, that approval can become the extraction tool. Periodically reviewing and revoking approvals is not paranoia; it's basic hygiene.

The interface risk is also underappreciated. You don't interact with "the blockchain" in some abstract sense; you interact through a website, a wallet app, a browser, a device. A compromised front end can feed you a malicious transaction while the underlying contract remains "legit." Malware can swap a copied address in your clipboard for the attacker's address. A fake wallet can look real while siphoning secrets. In other words, the weakest link is often not cryptography; it's the meatspace around it.

Centralized platforms come with a different failure mode: you can do everything right and still be stuck. If a trading app freezes withdrawals, experiences a run, gets hacked, mismanages customer assets, or collapses under leverage, your personal operational excellence doesn't help much. The risk is less "did you click the wrong link?" and more "did you choose the wrong counterparty?" In finance, counterparty risk is the risk that the other side can't or won't perform. Crypto didn't invent it; it just made it visible to people who were previously protected by regulated intermediaries and deposit insurance.

Even the word "exchange" can be a trap. In traditional markets, exchanges and broker-dealers live in heavily regulated regimes with rules around customer protections, disclosures, and supervision. Online platforms in crypto can borrow the language of legitimacy without the structure underneath. A slick interface can create an illusion of oversight. The defense here is not cynicism; it's verification. Where is the company domiciled? What regulatory registrations does it claim, and can you verify them through official sources? What are the withdrawal policies in plain language? Does it publish meaningful information about custody, controls, and audits, or just marketing content?

A personal risk framework helps because it prevents you from relying on vibes when you're excited. One practical way to do this is to evaluate any crypto activity using four questions:

First: What can I lose if something goes wrong? Not in percentage terms-dollars. A 70% drawdown sounds abstract; "this would wipe out three months of expenses" is clarifying.

Second: What is the failure mode? Is the main risk price volatility, platform insolvency, key loss, phishing, smart contract exploits, or regulatory freeze? Different failure modes require different mitigations.

Third: What is my recovery path? If the answer is "there is none," the position size should shrink accordingly. If a platform has support and legal presence, there is at least a procedural path. If you send funds to the wrong chain or sign a malicious transaction, recovery is often a fairy tale.

Fourth: What must be true for this to be safe? This forces assumptions into daylight: "the dev team won't rug," "the contract is secure," "the bridge won't fail," "my device is clean," "the exchange is solvent." The longer that list, the more you are stacking correlated risks.

Once you think this way, the day-to-day safety habits stop feeling like chores and start feeling like basic controls. Use hardware wallets for meaningful amounts. Keep devices updated. Use a dedicated browser profile for crypto. Turn on multi-factor authentication (and prefer app-based authenticators over SMS when possible). Whitelist withdrawal addresses if your platform supports it. Test transfers with small amounts before moving large sums. Separate "hot" wallets (for experimentation) from "cold" storage (for savings). None of these are glamorous, which is exactly why they work. Criminals thrive on excitement and haste.

It also helps to be honest about the difference between *complicated* and *sophisticated*. Some scams are complicated on purpose: they bury you in jargon so you feel too embarrassed to ask questions. Sophistication, in finance, usually produces clarity: clear terms, clear risks, clear disclosures, clear governance. If the explanation of how you make money sounds like a magic trick-"AI bot," "guaranteed returns," "risk-free yield," "secret arbitrage"-the sophistication is probably in the scammer's targeting model, not in the product.

Finally, accept that the safest investor is not the person who can spot every scam. It's the person who has designed their behavior so that a scam needs multiple miracles to succeed. If you never share your seed phrase, a huge category of attacks dies immediately. If you don't click inbound links, another category dies. If you don't grant unlimited approvals casually, another category dies. If you don't treat anonymous DMs as financial advice, another category dies. Crypto safety is less about heroic intelligence and more about unromantic constraints.

"Don't get rugged" sounds like a meme, but it's actually a philosophy: assume incentives matter, assume failure is possible, and build habits that make your worst day survivable. In a world where money can move at the speed of software, prudence isn't old-fashioned. It's a competitive edge.

Chapter 6

Strategy: Building a Digital Asset Portfolio

Buying digital assets without a strategy is just gambling with extra steps. This chapter moves beyond the 'HODL' memes to treat digital assets as what they really are: venture-stage technology bets. We will translate traditional investment discipline-valuation, due diligence, and risk management-into the language of blockchain, helping you build a portfolio that can survive the volatility without wrecking your sleep.

6.1 Valuation: Pricing the Priceless

A share of stock comes with a built-in translator. Every quarter, the business converts messy reality—customers, competition, mistakes, breakthroughs—into a familiar dialect: revenue, margins, cash flow. A bond is even more blunt: you lend, you get paid, you count the payments. Most digital assets show up to the valuation party with no such translator. No income statement. No dividend policy. Often no legal claim on anything at all. And yet they trade, sometimes with a confidence that borders on theatrical.

That mismatch is the first clue. Valuing a token is less like valuing a company and more like valuing a *system*. You are pricing a set of rules (the software), a set of incentives (who gets rewarded for what), a set of constraints (what can never exceed the cap, what can be changed by governance), and a set of habits (what real people do with it when nobody is watching). The investor's job is to turn those into a story with numbers—and then to be suspicious of the story.

The second clue is that the market already *has* a valuation framework for hard-to-pin-down things. It just doesn't call it that. When venture capitalists price an early-stage network, they aren't discounting mature cash flows; they're betting on adoption curves, retention, and the possibility that a platform becomes unavoidable. When someone buys a prime piece of land in a city that's still growing, they're valuing location, constraints on supply, and future foot traffic. Tokens sit at the intersection of those two instincts: the network bet and the scarce-asset bet. Most of the confusion in crypto valuation comes from mixing them up mid-sentence.

Start with a simple, clarifying question: *what kind of thing is this token supposed to be?* Not in marketing language—in economic language.

Is it a money-like asset? Then your mind naturally goes toward scarcity, credibility, settlement utility, and monetization of trust.

Is it a consumable input? (A "gas" token, a token required to pay for computation, storage, or bandwidth.) Then you care about usage demand, fee markets, and whether demand is real or subsidized.

Is it a governance chip? Then you're pricing control: the ability to steer parameters, treasury spending, or future economics—and the risk that governance is theater because a few wallets hold the votes.

Is it a claim on cash flows? Some tokens are closer than people admit: they route protocol fees to holders, or they represent a stake in a validator set that earns fees. If cash truly reaches token holders in a durable way, you can bring out a discounted cash flow model with fewer apologies.

The trick is that many tokens pretend to be all of these at once. A project will say "utility" when it wants to avoid sounding like a security, "store of value" when it wants to sound inevitable, and "governance" when it wants to sound decentralized. Your valuation work begins when you force the token to pick a dominant role. If it can't, you're not looking at a multifaceted asset; you're looking at a narrative looking for a price.

Market cap is the headline; supply is the fine print

Crypto markets love a clean headline number: market capitalization. It's familiar. It looks like equity. It shows up on screeners and social posts. And it is, in a narrow sense, accurate: current price multiplied by current circulating supply.

But market cap is a snapshot of today's float, not a portrait of tomorrow's dilution.

Most tokens are not born fully distributed. They hatch. Teams, investors, foundations, liquidity incentives, ecosystem grants—all of that tends to unlock over time. The schedule may be transparent, it may be complicated, or it may be disguised as "community emissions" that quietly act like payroll. Either way, price discovery happens on the slice that is tradable *today*, even though the economic reality is determined by the supply that becomes tradable *later*.

That's why fully diluted valuation (FDV) matters. FDV is price multiplied by the maximum supply (or total supply, depending on how the token is defined). It is not a prophecy; it is a stress test. If the project succeeds, and if the market still likes the token, the market has to absorb the unlocked supply without collapsing the price. FDV asks a blunt question: *if every promised token were liquid at today's price, what would this network be "worth"?* The answer is often uncomfortably large, and that discomfort is the point.

FDV can be misused. Some tokens have burns, variable issuance, or supply that depends on governance decisions. Some have never committed to a true cap. In those cases, the honest move is to replace FDV with a supply scenario: "supply in 12 months, 36 months, and at maturity under reasonable assumptions." But the spirit is the same. You do not want to be the investor who buys a "cheap" market cap and later discovers it was cheap only because most of the supply was still behind the curtain.

A practical way to make this real is to treat token supply like a second income statement—a schedule of future pressure. Equity investors watch share count and dilution from stock-based compensation. Token investors should do the same, except the dilution may arrive faster and with fewer constraints.

Tokenomics is not a buzzword; it's the asset

Tokenomics sounds like branding until you translate it into normal financial language. It's simply *supply and demand mechanics written into code.* In equities, management can decide to issue shares, buy back shares, change dividend policy, raise prices, cut costs, or pivot the business. In many token networks, the most important "management decisions" are pre-committed: issuance rules, fee allocation, emissions, burns, staking rewards, slashing penalties. That's not a side detail. That *is* the asset.

The most useful mental shift is to stop thinking of tokenomics as a set of features and start thinking of it as a *term sheet you cannot renegotiate.* When you buy a token, you are agreeing to be diluted (or not), taxed by fees (or not), rewarded for staking (or not), and potentially exposed to governance changes that can rewrite the bargain. If you wouldn't invest in a startup without reading the cap table and the liquidation preferences, it's strange to invest in a token without reading the equivalent.

A good tokenomic design tends to have three qualities that feel almost old-fashioned:

1. *The token is needed for something that persists* (not a one-time gimmick, not a subsidized loop that breaks when incentives end).
2. *Value capture is explicit* (holders, stakers, or validators have a clear path to being compensated by real usage).

3. *The rules are credible* (either truly hard to change, or governed in a way that won't casually betray minority holders).

A bad tokenomic design is often recognizable by the smell of free lunch. If a protocol promises high "yield" while also promising that token holders won't be diluted, and also promises that fees are low, and also promises that users will come anyway, something is off. Finance is generous with innovation, but it is not generous with arithmetic.

The two biggest lies: ``yield" and ``revenue"

Tokens don't have earnings reports, but analysts still crave a familiar anchor. So the market improvises two substitutes: "yield" and "revenue." Both can be real. Both are frequently misleading.

Yield in crypto is often paid in the token itself. That matters. If a protocol issues 10% more tokens per year and distributes them to stakers, the staker sees more tokens, but the pie is also bigger. That is not free return; it is a reshuffling of ownership. In traditional finance terms, it's closer to a stock split combined with a transfer from non-participants to participants. The correct question isn't "what is the APY?" It's "what is the APY after dilution and after any change in token price driven by that dilution?"

Revenue is even trickier. Some dashboards call all fees "revenue." Others distinguish between fees paid by users, fees kept by liquidity providers, and fees routed to the protocol treasury. Some projects count token emissions as "incentives," which is polite language for printing. In equities, revenue comes from customers paying the company. In token networks, there are at least three different "customers" (users, speculators, and subsidy programs), and only one of them is a real customer in the old sense.

A surprisingly effective discipline is to demand a cash-flow map in plain English:

> Who pays? In what asset? Where does it go? Who can ultimately convert it into dollars without collapsing the system?

If the answer involves a chain of conversions that depends on perpetual price appreciation, you're not looking at revenue; you're looking at a circular economy.

A network can be valuable before it is profitable

There is a reason people reach for network metrics. Networks can be valuable in ways that look irrational through a corporate lens. A payment network, a messaging app, a marketplace, an operating system—these can be strategically priceless long before they are efficiently monetized. The value lives in coordination: many people doing the same thing in the same place, because everybody else is already there.

Digital asset networks try to monetize coordination directly. Instead of an app company capturing the upside through equity, the protocol tries to push value into a token. That's a bold idea. It means adoption can show up in places other than revenue—in active addresses, transaction counts, fees paid, developers building, assets locked, and even the stubborn persistence of a community that keeps showing up after the hype has moved on.

A network valuation mindset usually leans on some version of Metcalfe's Law, the notion that the value of a network scales with the square of its users (or, more carefully, with the number of potential connections). You do not need to believe the exponent is exactly 2 to find the intuition useful. What matters is the shape: network value accelerates as adoption crosses thresholds. Early growth looks meaningless; later growth looks inevitable.

The danger is obvious: network metrics are easy to inflate. A protocol can subsidize activity. It can airdrop tokens to create bursts of addresses. It can wash volume through incentivized trades. It can count bots as users. It can stuff a metric the way a company can stuff a channel. If you're going to use network activity as the new P/E, you need the crypto equivalent of accounting skepticism.

A useful question for any activity metric is: *what would happen to this number if incentives were turned off for 90 days?* Some networks would sag and then recover. Others would go silent. That difference is not cosmetic; it is the difference between demand and bribery.

The Equation of Exchange: a finance old-timer that fits surprisingly well

One of the most practical bridges between monetary economics and token valuation is the Equation of Exchange:

$$MV = PQ$$

In plain English: money supply times velocity equals price level times quantity of economic activity. You can adapt it to tokens by treating:

- M: the effective token supply available for spending (not necessarily total supply, especially if large portions are staked or locked)
- V: how frequently the token changes hands (velocity)
- PQ: the dollar value of economic activity the token is used to settle (fees paid, payments made, value transferred—depending on the token's role)

This framework forces two uncomfortable realizations.

First, high velocity can be bearish for a token that is supposed to store value. If everyone rushes to sell the token the moment they receive it (for example, because it is used as a transient gas token), then the token has to work harder to maintain price. Lower velocity—people holding because the token is credibly useful or credibly scarce—supports higher valuation for a given level of economic activity.

Second, apparent "usage" can be a mirage if it doesn't require holding. If a token can be instantly swapped in and out at the moment of payment, usage does not necessarily create durable demand for the token itself. This is the difference between needing to *own* a thing and merely needing to *touch* it for a millisecond. Many token models assume ownership demand when the user experience actually creates touch demand.

The Equation of Exchange won't hand you a single correct price. What it does give you is a disciplined way to ask whether a valuation is internally consistent. If a token's market value implies a level of economic activity that would require the entire planet to be using it daily, you don't need a PhD to notice the strain.

Valuing fee-generating protocols: when DCF stops being heresy

The moment a protocol reliably captures fees and routes them to token holders (or to stakers in a way that is economically equivalent), traditional finance gets to come back into the room. Discounted cash flow (DCF) analysis becomes possible, not because crypto suddenly becomes normal, but because cash flow is a universal language.

But DCF in digital assets has special booby traps:

- *Fees are not profits.* Protocols often pay out most fees to liquidity providers, validators, or other participants. What matters is what remains for the token's economic claim.
- *Fee stability is not guaranteed.* Crypto fees are famously cyclical, driven by bull markets, memetic mania, and temporary congestion. Using peak fees as "run-rate" is the crypto version of valuing an oil company at the top of an oil spike.
- *Governance can change the split.* A protocol that routes fees to token holders today can vote tomorrow to redirect them to a treasury, to growth incentives, or to another stakeholder group. In equities, boards have fiduciary duties; in token governance, incentives are more fluid.

- *Competition arrives as code.* If a competitor can fork your protocol, copy your interface, and bribe your users with incentives, your margins are not protected the way a regulated utility's margins are.

Still, if you can build a conservative fee model and a realistic path of value capture, DCF can be a powerful sanity check. Even rough cash flow math has a moral quality in a market that loves stories: it forces you to state assumptions and accept consequences.

A companion approach is relative valuation: price-to-fees or market cap-to-protocol-revenue multiples (carefully defined). This is tempting because it feels fast. The danger is that it rewards the protocols most aggressively subsidizing activity. If fees are high because users are being paid to transact, the multiple is meaningless. Relative multiples work best when comparing mature protocols with similar subsidy intensity and similar value capture mechanics—which is rarer than people admit.

Bitcoin and the lesson of competing models

Bitcoin is a useful case study not because it makes valuation easy, but because it makes valuation *honest*. Even after more than a decade of trading history, there is no single accepted model that everyone agrees on. People have tried total addressable market (digital gold), stock-to-flow (scarcity), cost-of-production (mining economics), Metcalfe-style network models (adoption), and hybrids. Each captures something true; each also has failure modes.

That plurality is not a weakness to be embarrassed about. It's a signal that the asset does not fit neatly into a single category. Bitcoin behaves like a monetary asset, a risk-on speculative instrument, a geopolitical hedge, and a technology adoption story at different times. The correct response isn't to pick your favorite model and declare victory. The correct response is to triangulate: if your price target only works under one fragile assumption, you don't have valuation; you have belief.

Triangulation is the adult posture in digital asset valuation. Instead of pretending precision, you build a valuation range that is sensitive to key variables and you track whether reality is moving toward or away from the assumptions that justify the price.

A valuation checklist that doesn't pretend to be perfect

A practical valuation process for a token can be built from five buckets. None is sufficient alone; together they reduce the chance that you're paying luxury prices for a commodity.

1) **Supply schedule and dilution pressure.** Map circulating supply today, unlocked supply over time, emissions, burns, and any discretionary minting. Look for cliffs. Look for large allocations to insiders that become liquid during likely hype windows. Pay attention to whether emissions are funding real security and growth or simply manufacturing headline yields.

2) **Demand drivers with teeth.** Identify what forces someone to hold the token rather than rent it briefly. Staking requirements, collateral usage, governance that matters, fee discounts that are nontrivial—these can create ownership demand. Pure "utility" that can be instantly swapped for at the point of use often creates weaker demand than people assume.

3) **Network activity quality.** Prefer metrics that are expensive to fake. Fees paid in non-inflationary assets tend to be harder to counterfeit than raw transaction counts. Retention (do users come back?) matters more than bursts. Developer activity can matter, but only if it translates into products people use.

4) **Value capture mechanics.** Trace the path from usage to economics. If fees exist, who receives them? If a token is staked, what is the reward source: real fees or inflation? If tokens are burned, is

the burn material relative to issuance? If governance can change the rules, who controls governance?

5) **Market structure and reflexivity.** Liquidity, leverage, and narrative cycles can dominate fundamentals for long periods. Thin liquidity can make a token look "cheap" until you try to buy size. Incentive programs can make adoption look "real" until they end. The goal is not to eliminate reflexivity—it's to avoid mistaking it for a business model.

This checklist won't give you a single number, but it will give you a map of fragility. In digital assets, fragility is often a better predictor of future returns than elegance.

Pricing the option, not the certainty

Many tokens are priced like certainty when they are, in economic reality, options. An option is not an insult; it's a category. You pay for the right to benefit from a future outcome, with the understanding that the outcome may not arrive. Early-stage networks, new smart contract platforms, governance tokens for protocols still searching for product-market fit—these are option-like. They can go to zero. They can also become a piece of infrastructure that later looks obvious in hindsight.

Option-like assets should not be valued as if they were mature cash-flow streams. They should be valued with humility: lower position sizes, wider valuation ranges, and a willingness to change your mind as data arrives.

The uncomfortable flip side is that markets often price these options using momentum and narrative rather than sober probability. That creates opportunity, but it also creates traps. If your entire valuation case depends on "it will become the standard," you are not doing analysis; you are auditioning for a marketing deck.

A better way to phrase the bet is conditional: *if* adoption continues, *if* dilution is manageable, *if* value capture remains credible, *if* competition doesn't compress fees to dust, *then* this valuation could be justified. And then you ask: what do I need to see on-chain, in governance, and in user behavior over the next year for those *ifs* to remain intact?

That question produces an investment process, not just an entry price.

The most underrated metric: credibility

Traditional valuation models assume an institutional environment: audited financials, enforcement, fiduciary duties, mature disclosure regimes. Digital assets often operate with looser guardrails. That makes one variable unusually valuable: credibility. Not vibes—credible commitment.

Credibility shows up in places that feel almost mundane:

- Does the protocol do what it says under stress, or does it pause, patch, and rewrite rules when the market turns ugly?
- Are supply changes rare and hard-fought, or frequent and convenient?
- Are insiders transparent about wallets and vesting, or do tokens drift quietly onto exchanges?
- Is governance dispersed in practice, or concentrated behind a "community" label?

Credibility is valuation support. In a world where the asset is partly a set of promises encoded in software and social consensus, the credibility of those promises is not soft. It's structural.

When credibility is high, investors can accept lower cash flow today because the rules are expected to hold long enough for the economics to mature. When credibility is low, even attractive tokenomics can behave like a mirage, because the moment the bargain becomes valuable, someone will be tempted to change it.

A closing way to think: the token is a scoreboard

A token price is a scoreboard, not the game. It updates faster than fundamentals, reacts to rumors, and occasionally celebrates the wrong team. Valuation is the discipline of asking what the scoreboard is supposed to be measuring.

For a money-like token, it measures trust, scarcity, and settlement utility.

For a smart contract platform token, it measures demand for blockspace, security economics, and the credibility of monetary policy.

For a DeFi governance token, it measures the market's belief that governance matters and that fees will flow to the token rather than around it.

Those are measurable ideas. Not perfectly measurable—but measurable enough to build a range, track the inputs, and notice when the narrative starts running ahead of the numbers.

Pricing the priceless is not about pretending you can compute an intrinsic value down to the second decimal place. It's about refusing to buy a number without understanding the machinery underneath it. In a market built from code, the machinery is the investment.

6.2 The DYOR (Do Your Own Research) Field Guide

"DYOR" is crypto's version of "read the fine print," except the fine print is usually a Discord channel, a GitHub repo, a token distribution chart someone cropped out of a PDF, and a founder who posts like a motivational speaker. The slogan sounds empowering. In practice it often means: if something breaks, it's your fault for trusting anyone. That's not research; that's liability laundering.

Real research is less romantic and more effective. It treats every new token like an unfamiliar financial product sold by an unfamiliar institution in a lightly regulated mall. You assume three things until proven otherwise: the incentives are not aligned, the disclosures are incomplete, and the happy-path demo avoids the hardest questions. The goal is not to become paranoid. The goal is to become *boring*—the kind of boring that keeps you solvent.

The fastest way to lose money in digital assets is to confuse *interesting technology* with *investable asset*. Those are different species. Interesting technology can be brilliant and still be a terrible token. Investable assets can be dull and still be lucrative. DYOR is the filtration process that keeps your portfolio from becoming a museum of clever failures.

A funnel, not a rabbit hole

Research fails most often because it has no stopping rules. People start with a token, open twenty tabs, learn three new acronyms, and end up "investing" simply to justify the time already spent. A funnel is the opposite: you begin with a wide net and you reject quickly. Only a few candidates earn deeper work.

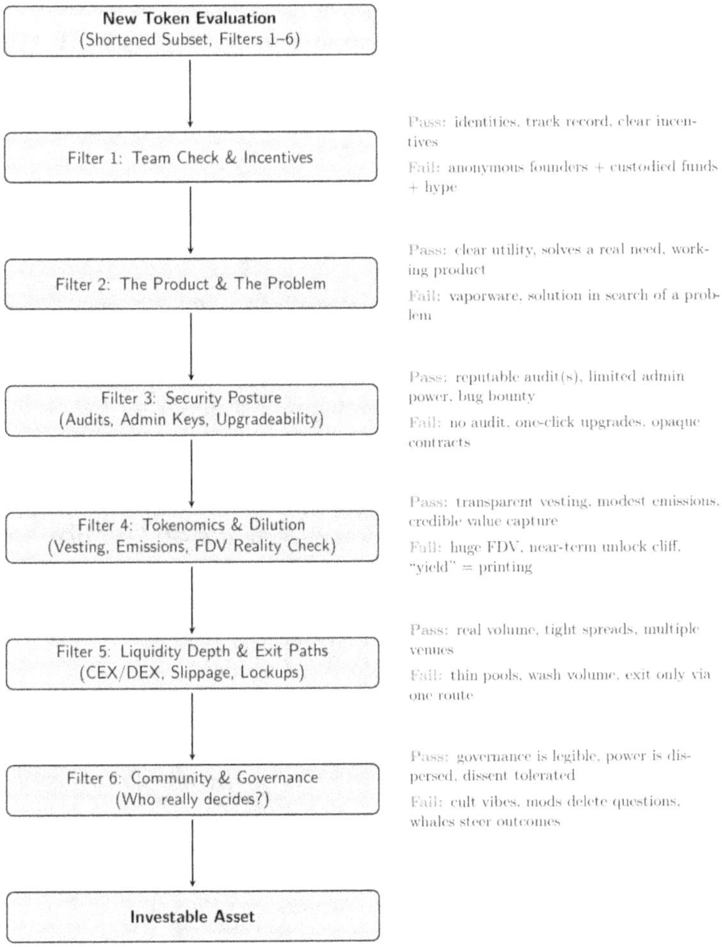

A funnel also forces a psychological shift: you're not trying to prove the token is good. You're trying to discover whether it deserves the scarce resource of your attention. Many tokens fail early. That's a feature, not a tragedy. The market manufactures more "opportunities" than any serious investor can process; your edge is refusing most of them quickly and calmly.

What follows is a practical checklist built to be used under real-world conditions: limited time, imperfect information, and a market that rewards confidence more than accuracy. The order matters. Early filters are cheap and decisive. Later filters are deeper and only worth doing when the basics survive.

Filter 0: The 60-second sanity check (before you get emotionally attached)

Before reading a whitepaper or watching a founder interview, look for the fastest disqualifiers. These are the investing equivalent of noticing a restaurant has a "C" health grade posted on the door.

Does the pitch rely on a single magical sentence? "It's like Amazon, but decentralized." "It's AI plus blockchain." "It's a stablecoin backed by vibes." If the explanation depends on you nodding politely rather than understanding, you're being sold to.

Is the token a solution in search of a problem? A protocol can be useful without a token. A token must justify its existence. If you remove the token and the product still works, the token is probably a fundraising mechanism with a marketing budget.

Is the upside described as inevitable? "Programmed to go up" is not analysis; it is mood lighting. Any asset pitched as destiny deserves extra skepticism.

Is there a clear way to lose money described in plain English? Good teams are oddly comfortable explaining risks. Bad teams treat risk disclosure as an act of disloyalty.

This quick check doesn't prove anything is safe. It only prevents the most common failure mode in crypto: falling in love with a narrative and then doing "research" as a form of rationalization.

Filter 1: Team check---the humans behind the code

Crypto culture sometimes romanticizes anonymity. In reality, anonymity is a risk multiplier the moment a project touches user funds, promises yield, or controls a treasury. You don't need founders to be famous. You do need them to be *accountable* in some way that survives market stress.

Identity and reputation: are these people real in the ways that matter? A real name isn't enough; plenty of scammers use real names. Look for continuity: long-standing profiles, consistent work history, credible references, past projects that still exist, and interactions with other reputable builders. A founder who appears fully formed on social media six weeks ago is not early-stage; it's freshly manufactured.

Track record: what did they build when nobody was watching? Anyone can hire a designer for a slick site. Fewer can point to code shipped, products maintained, and communities supported through boring periods. The most informative signal is often unglamorous: a history of fixing bugs publicly, responding to issues, and making incremental improvements.

Incentives: who gets rich, when, and for doing what? This is the cap table question, translated into tokens. If insiders unlock a large percentage of supply early, they may be rationally incentivized to maximize hype rather than durability. If the project pays itself mainly through token emissions, ask whether that's funding real work or simply paying salaries in a currency it prints.

Communication style: do they answer hard questions without performing? Watch how the team responds to skeptical questions about token allocation, admin keys, and governance power. Defensive sarcasm, moderator censorship, or vague appeals to "trust the process" are not personality quirks; they are governance previews.

A useful, slightly cynical rule: if the team's primary skill is *audience building*, treat the token as a media product until proven otherwise.

Filter 2: The product and the problem---does it earn its existence?

Some of the best scams in history were technically competent. A well-built product can still be a bad investment if it solves a small problem, or if the token is irrelevant to the solution. So this filter is not about admiration; it's about necessity.

Problem clarity: can you explain the problem without using the project's branding? If the problem can only be described using the project's invented vocabulary, it may not be a problem; it may be marketing.

User story: who uses it on a normal Tuesday? Not during a bull market, not during an airdrop, not during a liquidity incentive campaign. A normal Tuesday. Name the user: a trader, a developer, a business, a gamer, a remittance sender. Then ask what they do, why they do it here, and why they can't easily do it elsewhere.

Token necessity: what breaks if the token goes away? A token can be necessary as a security mechanism (staking), as a payment unit (fees), as a scarce resource (blockspace), or as governance (parameter control). But necessity must be structural, not decorative. "We need the token for the ecosystem" is the crypto version of "synergy."

Competitive moat: what stops copycats? Code is easy to copy. Communities can be bribed. If the moat is simply "first mover," you're buying a lottery ticket on brand dominance. That can work, but call it what it is.

When the product passes this filter, you should feel a subtle shift: the token starts to look less like a collectible and more like a component in an economic machine. That's where valuation becomes possible.

Filter 3: Security posture---because the best token is worthless if it can be drained

Traditional finance has its own failure modes, but it rarely includes "someone exploited a math bug and emptied the vault in nine minutes." Crypto does. The security filter is not optional; it's the price of admission.

Audits: not "did they get audited," but "what does the audit actually mean?" An audit is a point-in-time review of a codebase by a particular firm under a particular scope. It is not insurance. It is not a guarantee. Read the summary and look for blunt language: unresolved issues, "acknowledged" risks, excluded components, reliance on privileged roles. A project that brags about an audit but won't link it is telling you something.

Admin keys and upgradeability: who can change the rules overnight? Many protocols are upgradable. That can be reasonable early on. It is also an attack surface and a governance risk. If a small group can upgrade contracts without meaningful delay or oversight, then the protocol is not a neutral machine; it's a company wearing a protocol costume. That may still be investable, but you should price it like a centralized risk.

Bug bounties and incident history: do they pay for scrutiny or only for hype? A meaningful bug bounty program is a sign the team expects to be attacked and is willing to pay for defense. Incident history matters too. One exploit doesn't automatically disqualify a project; denial, secrecy, and sloppy remediation do.

Operational dependencies: what else must work for this to be safe? Oracles, bridges, multisigs, sequencers, custodians, front-end websites, RPC providers—these are the plumbing. A protocol can have perfect smart contracts and still fail because a dependency fails. The research habit here is simple: list the dependencies and ask which ones are centralized choke points.

If this sounds like work, good. You're doing what most market participants avoid, which is exactly why it has value.

Filter 4: Tokenomics under a microscope---where ``cheap" often means ``not yet diluted"

Many investors do tokenomics backwards: they start with price charts and then search for tokenomics that justify the price action. Better to treat tokenomics like a contract and ask whether you'd sign it if the token were not already pumping.

Supply schedule: when does the selling pressure arrive? Identify total supply, circulating supply, and the unlock calendar. Look for cliffs: sudden releases that can swamp liquidity. Watch for categories like "ecosystem incentives" that sound benevolent but function as ongoing issuance.

Fully diluted valuation (FDV): the uncomfortable mirror. Market cap is what's trading. FDV is what's promised. A low market cap with a high FDV is not necessarily undervalued; it may be *pre-diluted* in waiting. FDV also helps compare projects that have very different circulating supply percentages. It's not perfect, but it prevents a common rookie mistake: believing a small float implies a small network.

Emissions vs. real demand: are you being paid in printed coupons? High staking or farming rewards can be funded by inflation. Inflation can be justified if it purchases security or adoption that later becomes self-sustaining. It can also be a treadmill that requires constant new buyers to absorb new supply. The diagnostic question is brutal and effective: *what percentage of demand is organic if incentives drop sharply?*

Value capture: who benefits when the protocol succeeds? Some protocols generate impressive fee volume but route the economics away from the token. Others route value to a treasury controlled by governance that is effectively centralized. Others truly share economics with stakers or token holders. Trace the flows like an accountant: fees in, rewards out, burns, treasury accrual, and who controls the treasury.

Tokenomics is where optimism meets math. Markets can ignore math for a while. They rarely ignore it forever.

Filter 5: Liquidity depth and exit paths---the part nobody screenshares

A token is not an investment if you can't enter and exit without donating a chunk of your capital to slippage, spreads, and fees. Liquidity is also where manipulation hides. A chart can look healthy while the order book is hollow.

Where does it trade, really? A token "listed" on an exchange can still have thin real liquidity. Look across venues. Compare volumes. If most volume is on one obscure exchange or one shallow DEX pool, your exit plan is fragile.

Slippage reality check: how much can you buy or sell without moving the market? Try quoting a trade size that matters to you and see what the price impact is. Investors often fantasize about liquidity. The market charges tuition for that fantasy.

Concentration: who holds the supply? If a small number of wallets hold most tokens, liquidity can vanish the moment one of them decides to leave. Concentration also affects governance. A token with decentralized marketing and centralized holdings is decentralized in the same way a company picnic is "community-run."

Lockups and transfer restrictions: can you move the asset when you need to? Some tokens have vesting for insiders (fine), but others impose restrictions or have mechanisms that effectively gate transfers. These details matter most during stress, which is exactly when you will care.

A solid liquidity profile doesn't make a token good. It makes it tradable. And tradability is the difference between an investment thesis and a hostage situation.

Filter 6: Community and governance---a window into future betrayal

Community is often sold as a moat. Sometimes it is. More often it is a sentiment engine that amplifies whatever the largest holders want amplified. Governance is similar: it can be meaningful power, or it can be a ritual that gives retail the feeling of influence.

Where does real power live? Look for the actual levers: admin keys, multisigs, foundation control, upgrade permissions, treasury signers, and validator concentration. Voting tokens can be widely distributed while execution power remains tightly held. If the "community" can vote but cannot enforce, governance is theater.

How does dissent get treated? Healthy communities handle skepticism like a grown-up dinner party: they answer, debate, and sometimes admit uncertainty. Unhealthy communities handle skepticism like a cult: they shame, ban, and accuse. The latter tends to end the same way, because reality eventually becomes a dissenting voice.

Governance participation: who shows up? If proposals pass with tiny turnout, power is concentrated among those who bother to vote (often insiders and whales). Low participation also signals that the token is not held for governance; it's held for price appreciation.

Treasury behavior: is money spent like an institution or like a late-night shopping spree? Treasuries are where projects reveal maturity. Grants, audits, runway management, transparent reporting—these are good signs. Random sponsorships, influencer payments, and unexplained transfers are not "growth"; they're leakage.

Governance risk is not abstract. It directly affects valuation because it affects credibility. A protocol that can change the rules easily should trade at a discount, just like a company with weak shareholder protections.

Filter 7: Regulatory and consumer-protection reality---what protections do you *not* have?

Many investors import assumptions from brokerage accounts into crypto accounts. That's how people end up shocked when a platform freezes withdrawals or collapses and their "balance" turns out to be an unsecured claim in bankruptcy. The research habit here is to ask, plainly: *what legal and operational protections exist if this intermediary fails?*

Intermediary status: who is the counterparty? If you buy through an exchange, lend through a platform, or hold through a custodian, you have counterparty risk. Some platforms are registered in some jurisdictions for some activities; many are not. Even when a firm is reputable, the structure matters. You are not just buying an asset; you are choosing a set of legal relationships.

Proof of reserves: comforting screenshot or meaningful assurance? "Proof of reserves" can show that assets exist at a point in time. It often does *not* show liabilities, risk controls, or whether customer assets are segregated. It is not the same as a full financial statement audit. Treat it as a data point, not a guarantee.

Recourse: what happens after theft? Crypto theft is often irreversible. That fact should change how you size risk and how you evaluate any project that handles user funds. A protocol can be "legit" and still be an attractive target. The absence of recourse is a valuation input, because it affects expected loss.

This isn't about predicting the exact regulatory outcome for a token. It's about refusing to assume guardrails that may not exist.

Filter 8: Scam pattern recognition---because criminals also ship product

Scams evolve. Many are no longer crude. They use professional design, real customer support scripts, and even convincing "dashboards" that show fake profits. Some of the most damaging scams don't steal immediately; they cultivate trust, encourage larger deposits, and then lock the door.

A few patterns are worth treating like smoke alarms:

The "too smooth" onboarding. A stranger offers to coach you, share a "strategy," or guide you to a platform you've never heard of. The platform looks polished. Your first small deposit "works." The dashboard shows gains. Then the pressure begins: deposit more to unlock withdrawals, pay taxes upfront, or "verify" with another payment. Real financial institutions do not require you to pay extra money to withdraw your own money.

Romance and social engineering. The highest-return skill in crypto crime is not coding; it's persuasion. If investment advice arrives wrapped in emotional intimacy or urgency, treat it as a red alert, not a quirky origin story.

Impersonation and fake authority. Scammers impersonate project admins, customer support, even well-known investors. They mimic usernames and copy logos. The defensive move is boring but powerful: verify through independent channels. Do not trust inbound messages.

Community pressure as a weapon. "Only fools ask that question." "If you sell, you don't deserve to be here." "The team is working hard; stop spreading FUD." When a community uses shame to silence risk discussion, it's not building resilience. It's building exit liquidity.

Scam defense is part of DYOR because the market doesn't separate "investment risk" from "criminal risk." Your portfolio experiences both as the same thing: loss.

A lightweight DYOR workflow you can actually repeat

The best checklist is the one you'll use when you're tired, busy, and tempted. A practical workflow looks like this:

Step 1: One-page memo. Write a short note answering: What is the token for? Who are the users? What creates demand? What could kill it? If you can't write this in plain language, you do not understand it well enough to invest.

Step 2: Source triangulation. For every key claim (fees, users, audits, token supply), confirm it in at least two independent places, one of which should be primary (docs, contracts, reputable audit report). Screenshots and influencer threads don't count as primary sources.

Step 3: Identify the single biggest fragility. Every project has one. It might be dilution, admin-key risk, regulatory dependence, thin liquidity, or a weak moat. Name it. Decide whether you're being compensated for bearing it. If you can't name it, you're not being paid for it—you're ignoring it.

Step 4: Decide your stopping rule. Either it passes and becomes a monitored candidate, or it fails and you move on. Research without decisions is just entertainment with better vocabulary.

A final note that sounds obvious and still saves people: if you feel rushed, you are being rushed. Markets create urgency; scammers manufacture it. The best opportunities in digital assets rarely require you to decide in the next hour. The worst ones almost always do.

6.3 Sizing the Bet: Allocation and Risk Management

If valuation is the question "what is it worth?", allocation is the question that actually decides your financial future: *how much are you willing to be wrong?* Most investors talk about conviction as if it's a virtue. Markets don't reward virtue. They reward survival. Position sizing is where survival gets negotiated in advance, while you're calm and the chart isn't screaming.

Digital assets have a special talent for turning reasonable people into maximalists. The price moves are so large that they compress time.

A stock that doubles in two years makes you feel smart; a token that doubles in two weeks makes you feel chosen. Then it halves in three days and you feel personally insulted. That emotional whiplash is not a side effect of crypto. It's a central feature of the asset class, and it's why sizing the bet matters more here than in almost any other corner of markets.

There's also a structural asymmetry hiding in plain sight: traditional portfolios are designed around assets with deep markets, mature custody, and familiar protections. Crypto exposure often comes with extra risks that don't show up on a price chart—exchange outages, withdrawal freezes, smart contract failures, governance surprises, key compromise, and the simple fact that "owning" the asset can mean very different things depending on where and how you hold it. Allocation is not just about volatility; it's about the full menu of ways an investment can go wrong.

Start with a brutally simple rule: don't size a theory like it's a paycheck

Many crypto narratives are compelling precisely because they are *theories* about the future: money becoming native to the internet, finance becoming programmable, ownership moving into tokens. Theories can be right and still be uninvestable at large size. The world is full of correct ideas that arrived early, got implemented poorly, or got monetized by someone else.

So begin with a constraint, not a forecast. A useful constraint sounds like this: *if this entire position went to zero, my life would not change in a meaningful way.* Not your mood. Not your dinner conversation. Your life.

That sentence offends people in bull markets. It also prevents the most common long-term mistake: turning a speculative exposure into a lifestyle dependency. When your future happiness requires the token chart to cooperate, you are no longer investing. You are renting emotional stability from a market that charges late fees.

The 1--5% rule: a seatbelt, not a prophecy

A common starting point is the "1–5% rule": allocate a small slice of a diversified portfolio to digital assets, typically in the low single digits. The number isn't sacred. The logic is. Small allocations can matter because crypto returns, historically, have been lopsided—when it works, it can work loudly. But small allocations also respect the reality that crypto can experience deep drawdowns and idiosyncratic blow-ups unrelated to the broader economy.

Think of that 1–5% as a seatbelt. A seatbelt doesn't guarantee you'll never crash. It changes the consequences when you do. It keeps a drawdown from becoming a life event.

The other hidden benefit is behavioral: small allocations reduce the chance that you will make forced, emotional decisions. Crypto markets are open 24/7. Humans are not. A position that is too large will eventually meet you at 2 a.m., when you're tired, doom-scrolling, and convinced you must act immediately. The only sustainable defense is to size the exposure so that you can afford to be patient.

Risk capacity vs. risk tolerance: the difference between math and personality

Investors often say they have "high risk tolerance." That statement is usually tested in exactly one way: a large loss arrives, and the investor discovers whether tolerance was real or aspirational.

It helps to split the concept into two pieces:

Risk capacity is math. It's your balance sheet and cash flow. Do you have stable income? Emergency savings? Near-term obligations (tuition, down payment, medical costs)? If you need the money soon, your capacity for volatility is low, even if you enjoy adrenaline.

Risk tolerance is personality and psychology. How do you respond to uncertainty? Can you watch a position drop 50% without changing your strategy? Do you sleep? Do you pick fights? Do you stop doing your actual job to stare at charts?

Crypto punishes people who confuse the two. You can have high tolerance and low capacity (you love risk, but you can't afford it). You can have high capacity and low tolerance (you can afford volatility, but it makes you miserable). Allocation should respect the lower of the two.

A good way to operationalize this is to pre-commit to a drawdown you can live with. If your crypto sleeve dropped 70% (a number that is not exotic in this market), what would that do to your net worth? If the answer is "it would derail my plans," you already have the sizing answer, and it's smaller than your enthusiasm.

Correlation: diversification's unreliable friend

Diversification is the polite word for not needing to be right about any single thing. Crypto is often pitched as a diversifier: something different from stocks and bonds. Sometimes it is. Sometimes it behaves like a high-beta risk asset that falls when everything else falls. Correlations change, especially during stress, which is when you care most.

A measured way to hold both truths is to treat digital assets as *potentially diversifying over long horizons, but not reliably protective in crises.* That framing supports a modest allocation rather than a heroic one. If you size crypto as if it will save your portfolio in a crash, you are building on a fragile assumption. If it diversifies sometimes, great; if it doesn't, you still survive.

Also, correlation is not the only diversification variable. Crypto introduces distinct operational risks. Two assets can be uncorrelated on price and still be tightly linked through infrastructure: the same exchange, the same stablecoin rails, the same custodians, the same liquidity providers. In other words, you can diversify your tickers and still have a single point of failure.

The risk you don't see on the chart: operational drawdowns

Traditional finance mostly trains investors to think in price terms: volatility, drawdown, beta. Crypto adds a second axis: *operational drawdown*. That's a loss driven not by market price but by mechanics—the place you hold the asset fails, you lose access, you sign a bad transaction, a protocol is exploited, a bridge breaks, a wallet seed is compromised.

Operational risk changes sizing because it's lumpy. Markets usually give you time to react, even if they're cruel. Operational failures can be instantaneous and irreversible. That means allocation should be sized not just to withstand bad markets, but to withstand bad plumbing.

A practical approach is to assign your crypto sleeve an additional "haircut" in your mind. If you allocate 5% to digital assets, you might treat it as 4% in your planning assumptions, acknowledging that friction, fees, mistakes, and platform risk are part of the package. That mental haircut doesn't guarantee safety, but it keeps you from building a retirement plan on a best-case fantasy.

A core-and-satellite mindset without the jargon

Many investors get into trouble by building a portfolio like a playlist: whatever feels exciting this month. A more durable mental model is to separate money into two jobs:

- *Stability money*: the part designed to fund your life, protect you from surprises, and compound quietly. This is where boring is a feature.
- *Asymmetric money*: the part designed to take intelligent risk for outsize upside, with the explicit acceptance that it may underperform or even fail.

Crypto belongs, almost by definition, in the second bucket. Not because it's necessarily doomed, but because its return distribution is lopsided and its failure modes are wider. The key is that asymmetric money must remain a *slice,* not a takeover. If your speculative sleeve grows, that's good news—but it also triggers the need for rebalancing, because success is how crypto quietly becomes your entire financial plan.

Concentration risk: the quiet killer during bull markets

Bear markets hurt loudly. Bull markets hurt quietly, through concentration. The portfolio risk in a strong crypto rally is not that you lose money; it's that you make money and then become fragile.

This is a classic pattern: you start with 3% in digital assets. The market runs. That sleeve becomes 12% without you doing anything. Now your portfolio is meaningfully exposed to a single asset class with a history of deep drawdowns and operational blow-ups. You didn't make a decision to increase risk. The market made it for you.

That's why allocation is not a one-time choice. It's an ongoing relationship with drift.

Rebalancing: selling winners is a form of discipline, not betrayal

Rebalancing sounds like a dry housekeeping task until you experience it during mania. In a bull market, selling feels like heresy. Everyone around you has a price target that implies your current portfolio is just the early trailer. The temptation is to let it ride. Sometimes that works. It also turns a rational plan into an uncontrolled bet.

Rebalancing is the act of returning your portfolio to your chosen risk shape. It forces you to do something psychologically hard and financially sane: trim what went up and add to what lagged (or at least keep the laggard from being starved). It's a way of capturing gains without pretending you can pick the top.

A simple rebalancing rule that works because it's specific: set target ranges rather than exact targets. For example, if your crypto allocation target is 4%, you might allow it to float between 3% and 6%.

If it breaches 6%, you trim back to 4–5%. If it falls below 3%, you either top up (if your thesis is intact) or you reassess without panic. Ranges matter because crypto's volatility will otherwise force you to trade too often, paying fees and taxes while mistaking activity for prudence.

The deeper point is philosophical: rebalancing treats crypto like an asset class, not a religion. Religions demand unwavering faith. Portfolios demand maintenance.

Time-based vs. threshold-based rebalancing

There are two common ways to rebalance:

Time-based rebalancing happens on a schedule (monthly, quarterly, annually). It's easy to execute and reduces the chance that emotions dictate timing. In crypto, too-frequent time-based rebalancing can lead to overtrading; too-infrequent can allow dangerous drift.

Threshold-based rebalancing happens when allocations cross a predetermined boundary (for example, crypto exceeds 6% of the portfolio). This approach responds to volatility rather than the calendar. It often fits crypto better because the biggest risk is drift during sharp moves.

Many investors combine them: a quarterly review with a "rebalance if thresholds are breached" override. The main requirement is that the rule is written down before the market starts whispering that rules are for other people.

Liquidity planning: don't fund real life with unrealized gains

Crypto creates a specific kind of illusion: large unrealized gains that feel like cash. They are not cash until they are converted, and the conversion process can be interrupted by volatility, thin liquidity, tax consequences, exchange outages, or stablecoin rails under stress.

If you have any near-term cash needs, do not assume you can meet them by selling crypto at a good price on a convenient day. Build liquidity in advance through traditional cash reserves and conservative assets. Crypto can be part of long-term wealth building, but it is a poor emergency fund. Emergencies have a habit of arriving during market stress, precisely when crypto is most likely to be down and the infrastructure most likely to be strained.

A useful rule is to separate *funding* from *investing*. Funding needs should be met with assets designed to be liquid and stable. Investing assets can take risk. Mixing the two is how people become forced sellers.

Leverage: the fastest way to turn volatility into catastrophe

Crypto makes leverage feel casual: perpetual futures, margin, borrowing against collateral, recursive lending loops. The interface is clean. The liquidation engine is merciless.

Leverage is not just "more exposure." It changes the shape of outcomes. It introduces path dependency: the route matters, not just the destination. An asset can recover over a year, and you can still be liquidated on the way down in week two. Many investors discover this too late, because they confuse long-term conviction with short-term survivability.

If the goal is to hold digital assets for long-term upside, leverage is usually the wrong tool. It converts a long-duration thesis into a short-duration bet with hard triggers. You can be right and still lose. That is not a sophisticated risk; it is an avoidable one.

A practical ``crypto risk budget" you can live with

One way to make allocation concrete is to define a risk budget for crypto—a maximum acceptable loss in dollars, not percentages. For example: "I can accept losing $10,000 on crypto exposure without changing my life plans." Then you work backwards: if you expect crypto drawdowns can easily reach 60–80%, your position size should be small enough that a severe drawdown stays within that dollar budget.

This approach has two advantages. First, it anchors you to reality rather than returns. Second, it scales as your wealth changes. A $10,000 loss might be nothing for one investor and devastating for another. Percentages hide that difference; dollars don't.

When to increase size (and when not to)

Increasing a crypto allocation should feel less like falling in love and more like underwriting. Better reasons include:

- You have increased your emergency reserves and reduced near-term liabilities, raising risk capacity.
- Your research confidence has improved because the token's economics, security posture, and governance have matured.
- Liquidity has deepened, reducing exit fragility.
- You are rebalancing into a drawdown under a pre-committed rule, not chasing momentum.

Bad reasons are usually emotional disguises: fear of missing out, social proof, the belief that past returns guarantee future ones, or the feeling that you "deserve" a bigger position because you discovered it early.

The market does not pay people because they deserve it. It pays them because they managed risk while others lost their heads.

The quiet virtue: staying in the game

The most valuable skill in digital asset investing is not prediction. It's endurance. Endurance comes from sizing bets that let you hold through volatility without becoming reckless, and from rebalancing rules that stop success from turning into fragility.

A well-sized crypto allocation can add spice to a portfolio without setting the kitchen on fire. It can participate in upside, teach you about a new financial architecture, and potentially improve long-term returns. But only if it remains what it is: a high-variance exposure with unusual operational risk, managed with the kind of discipline that feels almost unfashionable in a market built on excitement.

The irony is that the investors who last the longest in crypto often look the least like crypto investors. They write rules down. They take profits without apology. They keep boring money boring. And when the market inevitably offers them a chance to act impulsively, they let it pass, because they already decided who they wanted to be.

6.4 Making Assets Work: Yield and DeFi

"Passive income" is one of finance's most abused phrases, and crypto found a way to make it even more slippery. In traditional markets, yield usually means something concrete: interest paid by a borrower, a coupon paid by an issuer, a dividend paid out of earnings. In DeFi, yield can mean those things—or it can mean you're being handed newly minted tokens as a reward for showing up. Both can look identical in an app: a glowing APY number and a button that says *Stake.*

The first mental adjustment is to treat yield as a *price tag on risk.* Nobody in DeFi pays you to be safe. If you're earning 18% on a stablecoin, the question is not "how do I get in?" The question is "what nightmare am I underwriting that the market demands 18% to tolerate?" DeFi is brilliant at packaging risk into something that feels like a savings account. Your job is to unpackage it.

Another adjustment is to remember that DeFi doesn't abolish the old laws of finance; it just writes them into software. There is still credit risk (the borrower may not repay), market risk (prices move), liquidity risk (you can't exit without pain), operational risk (systems break), and governance risk (rules change). The difference is that

DeFi often makes these risks faster, more mechanical, and less forgiving. If something goes wrong, there may be no customer support line, no regulator to call, and no "we'll reverse that transaction for you." The code does what it does. You live with it.

A working definition: real yield vs. printed yield

Before getting into specific strategies, it helps to sort yields into two buckets that behave very differently over time.

Real yield is paid out of economic activity: user fees, spread income, borrowing costs, liquidation penalties, or other cash-like flows generated because people actually used the system. Real yield can be volatile, cyclical, and competitive, but it has a grounding in demand.

Printed yield is paid from token issuance: the protocol inflates the token supply and hands new tokens to participants. This can be a rational bootstrapping tool (subsidize early adoption, pay for security, reward liquidity). It can also be a slow-motion transfer from late buyers to early recipients. Printed yield is not automatically bad; it's simply not the same thing as income.

If you want one sentence to keep you from being seduced by an APY banner, use this: *yield paid in something the protocol can print is not yield; it's dilution with better marketing*. Sometimes dilution is worth it. Sometimes it's the whole business model.

The practical test is straightforward: *if emissions went to zero tomorrow, would any meaningful yield remain?* If the answer is no, you are not earning income from a productive system. You are being paid to hold exposure.

Staking: getting paid to be the network's bouncer

Staking is often presented as the safe, grown-up way to earn in crypto. Sometimes it is. Conceptually, staking is simple: you lock up tokens to help secure a proof-of-stake network. In return, you receive rewards. Those rewards come from two places: (1) newly issued tokens (inflation) and (2) transaction fees paid by users. In many networks, inflation is the dominant component, especially in earlier stages.

The good news is that staking yield is not magic. It is compensation for doing a job: contributing to consensus and, in some designs, taking on the risk of penalties.

The bad news is that the job comes with risks that don't show up in the headline APY:

Slashing risk (penalty risk). Many proof-of-stake systems can penalize validators for misbehavior or serious downtime. If you stake through a validator, your return depends on that validator's competence and integrity. You are outsourcing an operational function, not buying a bond coupon.

Lock-up and liquidity risk. Some staking requires unbonding periods. That means you cannot instantly exit when markets move. You are accepting illiquidity in exchange for yield, which is a normal financial trade-off, but it should be priced consciously.

Depeg risk in liquid staking. Liquid staking tokens (which represent a claim on staked assets) add convenience: you can keep your position liquid and often use it in DeFi. They also add a new layer of risk: the token can trade below the value of the underlying staked asset during stress. That discount can be small, or it can widen sharply when everyone wants the exit at once. Liquidity has a habit of disappearing exactly when it is most precious.

Smart contract and dependency risk. Staking through a protocol wrapper introduces smart contract risk and dependency risk. Your exposure is no longer just "the network." It's the network plus the staking protocol plus the liquidity venue plus the oracle infrastructure that tells the world what things are worth. In DeFi, the yield often comes from stacking dependencies. The risk does too.

A helpful way to value staking yield is to treat it like a blended rate: part inflation transfer, part fee income, minus costs and risks. If you can't identify the sources of the yield, you can't evaluate whether it's sustainable.

Lending: yield that looks familiar, until the liquidation engine turns on

Crypto lending feels intuitively comfortable because the word "lending" carries centuries of financial meaning. Someone borrows. Someone lends. Interest is paid. In DeFi, the mechanics are often over-collateralized and automated. Borrowers post collateral worth more than the loan, and the protocol enforces safety through liquidation: if collateral value drops too much, it gets sold.

That liquidation feature is the heart of the system. It makes many DeFi lending protocols less like unsecured bank lending and more like a margin engine that is always on. The risk doesn't disappear; it moves around.

For lenders: the yield is compensation for system-wide stress. Lending yields usually rise when demand to borrow rises (often in bull markets) and can fall when demand dries up. Lenders are exposed to smart contract risk and to "tail events" where liquidation systems are overwhelmed by rapid price moves, oracle failures, or chain congestion.

For borrowers: the cost is not just interest; it is the risk of forced selling. Borrowing against crypto can look like a clever way to get liquidity without selling (and potentially without triggering taxable events, depending on jurisdiction and structure). The trap is that the loan is not patient. If the collateral falls and your health factor deteriorates, the protocol can liquidate you quickly. It does not negotiate. It does not care that you were "right long term." It cares about collateral coverage now.

The key risk concept here is *liquidation risk*: the possibility that volatility triggers an automatic sale of your collateral at the worst possible moment. In many lending systems, a small change in price can have a large impact on your safety buffer if you are levered. People often learn this the expensive way because the UI makes leverage feel like a slider, not a cliff.

A disciplined borrowing approach, when used at all, tends to have three features: conservative loan-to-value (a wide safety buffer), a clear plan for adding collateral or repaying during drawdowns, and a refusal to stack leverage on top of leverage. DeFi makes it easy to build elaborate recursive strategies. It does not make them robust.

Liquidity providing (LP): why ``earning fees" can still lose money

Liquidity providing is one of DeFi's most misunderstood yield sources because it looks like you're being paid for something wholesome: providing liquidity, improving markets, earning trading fees. The reality is more nuanced. In an automated market maker (AMM), you deposit two assets into a pool. Traders swap against the pool. You earn a share of fees. So far, so good.

Then comes the part most people don't internalize until it hurts: *impermanent loss*. Impermanent loss is the gap between (a) what you would have had if you simply held the assets and (b) what you have after providing liquidity and experiencing price movement. It happens because AMMs continuously rebalance your holdings as prices change. When one asset rises relative to the other, the pool sells it for you. You end up with more of the asset that went down and less of the one that went up. Fees can offset this. They can also fail to offset it.

Liquidity providing is not a savings product. It is a trading strategy with embedded rebalancing. In many cases, it is economically similar to being short volatility: you tend to do well when prices are range-bound and trading is active, and you tend to do poorly when prices trend strongly.

Concentrated liquidity (where you provide liquidity only within a price range) sharpens the blade. It can increase fee capture when the price stays in your range, but it increases sensitivity to price movement. It turns the position into something closer to a leveraged bet on where the price will *not* go. That can be lucrative. It can also be unforgiving.

The only honest way to evaluate LP yield is to compare total return, not fee APR. Total return includes fees *and* impermanent loss *and* any token incentives *and* the risk that the pool's code or surrounding infrastructure fails. If someone shows you only the fee APR, they are showing you the appetizer and hiding the calories.

Yield aggregation: when convenience becomes stacked fragility

DeFi has an entire layer of products that do something very appealing: they bundle yield opportunities and optimize them automatically. Yield aggregators can rebalance between pools, harvest rewards, and compound returns. The pitch is convenience and optimization.

The risk is that each layer adds dependencies. A base protocol can be safe and still be made risky by an aggregator that uses it incorrectly, integrates with fragile components, or introduces governance and admin-key vulnerabilities. Convenience in DeFi is rarely free; you pay in complexity, and complexity is where surprises live.

A simple research habit helps here: draw the dependency chain. Your funds flow through contracts A, B, and C, rely on oracle D, and exit through liquidity pool E. Each link is a potential failure point. If you cannot name the links, you are outsourcing understanding. Outsourcing understanding is fine for some things. It is expensive in DeFi.

The stablecoin yield temptation: ``it's just dollars" is where people get careless

The most psychologically dangerous yield in crypto is stablecoin yield, because it feels like a riskless rate. A stablecoin is supposed to be worth a dollar. Lending it at 12% feels like discovering a loophole in finance.

The problem is that stablecoin yield often concentrates multiple risks:

- *Depeg risk*: the stablecoin may not hold its peg under stress.
- *Counterparty or protocol risk*: the yield may rely on an intermediary or smart contract that can fail.
- *Liquidity risk*: exits can be gated, delayed, or expensive when everyone rushes at once.
- *Hidden leverage*: some yields are generated by borrowers taking leveraged positions elsewhere in the system.

The higher the stablecoin yield above prevailing short-term rates in traditional markets, the more aggressively you should hunt for the risk being smuggled in. Sometimes the yield is real and compensates for real risk. Sometimes it is a fragile structure held together by incentives and optimism. The dollar sign doesn't change the nature of the system.

A practical way to underwrite DeFi yield: ask ``who pays me?"

When a DeFi product offers yield, there are only a few sources of payment. The yield is ultimately coming from one or more of these:

- Borrowers paying interest
- Traders paying fees
- Liquidations paying penalties
- The protocol printing tokens (emissions)
- A treasury subsidizing usage (often funded by previous token sales)

Everything else is presentation.

Once you know who pays, you can ask whether that payer is durable. Borrowers may disappear when leverage is unattractive. Traders may vanish when volatility drops. Liquidations spike during crashes but can become chaotic and unprofitable if the system breaks. Emissions can persist as long as the protocol is willing to dilute holders. Subsidies last until the treasury gets tired.

This framing also makes it easier to avoid the most common DeFi trap: mistaking *redistribution* for *production*. Some yields are funded by genuine economic activity. Others are funded by transferring value from token buyers to token farmers. Both can persist for a while. Only one deserves to be treated like income.

Composability: the superpower that also creates chain reactions

DeFi's superpower is composability: the ability to stack protocols like Lego bricks. You can stake an asset, receive a liquid token, use it as collateral, borrow against it, provide liquidity, and farm incentives—all without asking permission. It's extraordinary engineering. It also creates a financial system where stress can propagate quickly because exposures are intertwined.

When yields look especially attractive, composability is often involved. A strategy may be earning from multiple sources at once, but it may also be taking the same risk multiple times in different costumes. For example, you might be exposed to the same underlying asset's price move through collateral, through LP positions, and through liquidation thresholds, all at once.

A smart risk-management habit is to look for *risk overlap*. If the strategy fails, what is the single underlying event that causes the cascade? Often it's a sharp drawdown in the collateral asset, an oracle failure, or a liquidity crunch. If one event can break the entire stack, the yield is paying you to sit on a trapdoor.

How to size yield strategies without fooling yourself

Yield strategies deserve even smaller sizing than directional "buy and hold" positions, because they add non-linear risks. A token held in cold storage can go down in price, but it won't be liquidated. A lending position can be.

A practical hierarchy of caution looks like this:

- Treat smart-contract-exposed yield as a separate risk sleeve inside crypto, not as a default use of all holdings.
- Avoid concentrating all yield exposure in one protocol, one stablecoin, or one chain.
- Assume that exits during stress will be worse than the app suggests.

Also, be honest about what you're optimizing. Many yield strategies are not just "earning." They are a way of taking additional risk in exchange for a return stream that *looks* smoother than price appreciation. Smoothness is seductive. It is also sometimes fake, because the risk is latent until it detonates.

The most useful mindset: treat DeFi yield like running a small financial business

If you want DeFi yield to be more than gambling, treat it like operating a small business. A business has revenue, costs, and risks. In DeFi:

- Revenue is fees, interest, or emissions.
- Costs are gas fees, slippage, hedging costs, and opportunity cost.
- Risks are smart contract exploits, liquidation, depegs, governance changes, and market regime shifts.

A real business owner tracks net profit, not gross receipts. A real DeFi participant should track net return after all costs and after realistic assumptions about tail risk. The discipline is the same even if the tools are different.

The payoff for doing this work is that DeFi can offer something genuinely valuable: a way to turn digital assets into productive capital. But productivity is not guaranteed by an APY label. It is earned by understanding the mechanism, refusing to confuse printed rewards with income, and sizing positions so that a broken contract does not become a broken life plan.

DeFi rewards curiosity, but it punishes carelessness. The investors who do best over time are rarely the ones chasing the highest yield. They're the ones who can explain, calmly and specifically, what they're being paid for—and what they would lose if the system stopped behaving.

Chapter 7

The Future: Enterprise and Capital Markets

While headlines often chase the volatility of cryptocurrencies, the quiet revolution is happening in the back offices of the world's largest banks. This isn't about speculation; it's about plumbing. We are witnessing an upgrade to the operating system of global finance-shifting from slow, paper-based legacy systems to instant, programmable digital workflows. In this chapter, we explore how major institutions are moving past the 'experimental' phase to rewire capital markets, automate compliance, and create a more efficient global economy.

7.1 Show Me the Money: The Business Case for Tokenization

Tokenization is one of those words that can mean something profound or absolutely nothing, depending on who's using it. In a boardroom slide deck, it often functions as a synonym for "modern." In production systems, it has a narrower, more brutal meaning: taking an asset, giving it a precise digital representation, and moving its lifecycle onto rails that can enforce rules and synchronize state across many parties. The business case lives or dies in that last clause. If tokenization merely *duplicates* what existing infrastructure already does—while keeping the same number of handoffs, checks, and reconciliations—it's not modernization. It's cosplay with a new database.

The honest question to ask is not "Does blockchain reduce costs?" but "*Which* costs?" Capital markets don't spend most of their money on the act of trading. They spend it on everything that happens because trading happened: matching records, fixing breaks, chasing signatures, calculating entitlements, updating multiple ledgers, managing collateral, satisfying compliance, and proving to auditors that the whole story is consistent. This is the unglamorous industry beneath the industry—post-trade plumbing, operations, controls, and supervision. Tokenization can compress that workload, but only when it replaces duplication with shared state and replaces human coordination with machine-enforced rules. That's a tight target. Miss it and you get a very expensive parallel universe.

Cost centers: where money actually leaks out of modern finance

The easiest way to spot "expensive tech theater" is to follow the headcount. A surprising amount of financial-infrastructure cost is labor dressed up as process: teams that reconcile positions between internal systems; teams that resolve settlement failures; teams that manage static data, corporate actions, and exceptions; teams that monitor eligibility rules and transfer restrictions; teams that prepare regulatory reports; teams that respond to audits with screenshots and spreadsheets. Many of these jobs exist because the financial system is a federation of separate ledgers. Each ledger is correct *locally*, but the system is correct only after everyone has compared notes.

Reconciliation is expensive because it's not a single task—it's a recurring tax on complexity. The same trade may be represented differently in a front-office order management system, a middle-office risk system, a back-office settlement system, a custodian's books, a clearing agent's books, and a central securities depository's records. Some differences are legitimate (different data models, different rounding, different time zones). Many are accidental (message formatting, stale reference data, manual adjustments). Every difference creates the possibility of a "break," and every break creates a mini-investigation. In a high-volume world, even a small break rate is a factory of tickets.

Tokenization's core promise is to collapse this federation into a smaller number of shared sources of truth—ideally one. That promise is not mystical. It's architectural. If the asset's state (ownership, encumbrances, eligibility, lifecycle events) is represented on a shared ledger that all relevant parties can read and update under clear rules, the reconciliation burden shrinks because there are fewer independent versions to reconcile. When the asset "moves," everyone sees it move. When it's pledged, everyone sees the pledge. When a coupon is paid, the entitlement logic is visible and deterministic. Less detective work, more bookkeeping.

But notice what's *not* being claimed: that tokenization eliminates operations. Operations doesn't disappear; it changes shape. Instead of staff manually bridging mismatched ledgers, you need staff designing controls, managing permissions, monitoring smart-contract behavior, handling exceptions that code can't resolve (and there will always be exceptions), and governing upgrades. If someone promises "zero operations," they are selling you a demo.

Time is a cost: why settlement latency is not just an inconvenience

Capital markets have a peculiar relationship with time. In many industries, time-to-complete is a customer-experience metric. In markets, time-to-complete becomes a risk metric and a funding metric. When a trade is agreed on Monday but settles on Wednesday (a typical $T + 2$ cadence), the system is carrying open exposures for two days. Someone is on the hook if the counterparty fails. Someone must fund positions while they are "in flight." Someone must post margin, maintain credit lines, and monitor limits. Time is not idle; time is a balance sheet.

This is where the tokenization pitch gets sharp: atomic settlement—delivery-versus-payment in one synchronized action—compresses exposure windows. If the cash leg and the asset leg settle together, there is no limbo where one side has delivered and the other is waiting. That reduction in counterparty risk can reduce the amount of collateral and credit support needed around the process. It can also reduce the operational scaffolding built to manage that limbo: confirmations, netting cycles, cut-off times, and the whole calendar of "processing days" that exist because systems need breathing room.

The catch is that atomic settlement is not a philosophical property of blockchains. It's an engineering property of *integrated settlement assets*. If your tokenized bond settles in a tokenized cash instrument that is accepted as final money by the participants—and ideally anchored to central bank money or a robust commercial-bank equivalent—then atomic settlement can be real. If your "cash leg" is a promise to wire dollars later, you've reinvented the same old two-step with a new user interface.

And even when atomic settlement is possible, "$T + 0$ everything" is not automatically optimal. Many market participants use settlement delay as a kind of built-in financing and workflow buffer. Netting processes reduce the number of movements by offsetting trades against each other; compressing settlement can increase gross settlement volumes and intraday liquidity needs. Tokenization changes the trade-offs. It doesn't make them vanish.

Programmability: the part people oversell and still underuse

"Programmability" gets marketed like magic: money that pays itself, assets that enforce rules, compliance that runs in the background. The reality is more practical and, ironically, more valuable. Programmability is the ability to attach *contingent actions* to an asset's lifecycle—actions that trigger when conditions are met. That could be a coupon payment, a margin call, a corporate action entitlement, a transfer restriction, a lock-up expiry, a collateral substitution, or a forced redemption event.

The cost savings come from moving conditional logic out of email chains and into deterministic execution. Today, many lifecycle events are executed through a choreography of instructions: a notice goes out, someone interprets it, someone keys it into a system, someone approves it, someone reconciles the result. Each step is a point of friction and a point of failure. Programmability doesn't eliminate the legal and economic meaning of the event; it makes the execution less ambiguous and more automatable.

But there's a reason this is underused: writing lifecycle logic is easy in the happy path and hard in the real world. Corporate actions can be messy. Bond terms can have edge cases. Funds have cutoff times, holiday calendars, and exceptions for failed payments. Regulators can impose freezes. Courts can order reversals. If the smart contract can't represent the messy parts, the system either becomes brittle or quietly reintroduces manual interventions—often via admin keys that are the on-chain equivalent of "someone in operations can fix it." Admin keys can be entirely legitimate (regulated finance needs controlled intervention), but they change the risk model. The business case must include governance and controls, not just automation.

The two ROI questions that matter: replacement vs. duplication

Every tokenization initiative should be forced to answer two uncomfortable questions:

First: what are we turning off? If the project does not retire a legacy system, retire a reconciliation process, retire a class of messaging, or materially reduce exception handling, then the savings story is mostly imaginary. A parallel ledger that must be reconciled back to the "real ledger" is not progress; it's additional work. Many early tokenization pilots fall into this trap because they start as "shadow mode"—running in parallel to prove safety. Shadow mode is sensible during experimentation. It becomes a permanent tax if you never exit it.

Second: who captures the savings? This is the coordination problem disguised as an ROI model. A bank might fund a tokenization project that reduces settlement failures across the street, but if the savings accrue mostly to custodians, clearing agents, or clients, the sponsor's business case looks weak. Conversely, a central infrastructure provider might capture efficiency benefits while pushing integration costs onto participants. Tokenization can create value and still fail commercially because incentives are misaligned. In markets, the best technology in the world cannot overcome "not my budget" politics.

A sober ROI model separates *private ROI* (benefits to the sponsor) from *system ROI* (benefits to the network). Tokenization is inherently networked. If you don't have a credible plan for network adoption—either by starting with a closed consortium, embedding into incumbent infrastructure, or offering a product so compelling others *choose* your rails—the savings will remain theoretical.

Collateral mobility: the sleeper use case that actually smells like money

If there is a tokenization use case with an almost unfairly strong business case, it's collateral. Not because collateral is glamorous, but because it is everywhere and chronically inefficient. The modern financial system runs on secured credit: repos, derivatives margining, securities lending, prime brokerage financing. Collateral is the bloodstream—and like the bloodstream, the value is not the blood itself but the ability to move it where it's needed, fast, without clotting.

Today, collateral is often trapped in silos. It sits at a custodian, pledged under one agreement, unavailable for another, or it can be moved only through processes that are slow and operationally heavy. Substitution—replacing one piece of pledged collateral with another—can be a paperwork marathon. Eligibility checks can be manual. Settlement timing can force participants to over-post "just in case." The result is a familiar inefficiency: the system holds more high-quality assets in reserve than it would need if collateral could flow more freely.

Tokenization can help by making collateral positions visible, composable, and programmable. A tokenized security can be locked (encumbered) in a way that is machine-readable: it can't be transferred unless conditions are met, but it can still be valued, monitored, and, in some architectures, rehypothecated under tightly defined rules. Haircuts, concentration limits, and eligibility criteria can be encoded as checks rather than enforced after the fact. When this works, it doesn't merely reduce processing cost; it reduces the amount of idle, unproductive "buffer collateral" institutions carry. That is real money: a balance-sheet efficiency, not just an operations efficiency.

The biggest caveat is legal and operational realism. Collateral is not just a technical lock; it's a legal claim under a specific agreement, with jurisdictional nuances and insolvency implications. A token that says "pledged" is not enough unless the legal framework treats that pledge as enforceable and the operational ecosystem (custody, valuation, dispute resolution) recognizes it. Tokenization can reduce friction, but it cannot wish away legal plumbing. The business case strengthens dramatically when tokenization is introduced in tandem with clear legal enforceability and with credible settlement assets.

When tokenization is just expensive theater

Theater has a recognizable smell: impressive demos, vague metrics, and lots of talk about "disintermediation" without naming which intermediaries are being removed. Finance has intermediaries for reasons—some are historical, some are regulatory, some are economic. Removing them can be smart, but only if their functions are replaced by something that actually works in production under stress.

A common theatrical pattern is "tokenization on the front end, traditional everything on the back end." A token is issued, investors buy it, and then the issuer's transfer agent maintains the real register off-chain. Corporate actions are handled off-chain. Settlement finality is off-chain. In that design, the token is more like a receipt than an asset. Receipts can be useful—receipts can improve distribution and transparency—but the ROI claim should be framed as product and distribution, not as "post-trade transformation."

Another theatrical pattern is "blockchain as a database upgrade." If a single institution controls all write access, all validation, and all governance—while using a distributed ledger internally—then the value proposition needs to be compared with conventional technology: replicated databases, event-sourcing architectures, permissioned data sharing, secure multiparty computation, even just better APIs. A blockchain may still be a reasonable choice for certain auditability and shared-state reasons, but it is not automatically cheaper than well-run centralized infrastructure. Distributed systems are famously costly to build and operate. The savings must come from *eliminating duplication across institutions*, not from adding new complexity inside one.

The most expensive theater, though, is the project that promises to "transform settlement" while leaving the settlement asset untouched. If the system still depends on end-of-day net settlement in existing rails, and if participants must still run the same risk and funding processes, then the main benefits of tokenization are blocked. What you get is a new messaging layer that still hands off to the old machinery. Sometimes that is a necessary transitional architecture. It is not, by itself, a cost revolution.

Integration costs: the part that eats your ROI while nobody is looking

Tokenization projects often underestimate integration cost because the demo looks clean. A token moves from Wallet A to Wallet B and a block explorer shows it happened. The demo does not show the integration to risk systems, accounting systems, client reporting, tax lots, NAV calculations, regulatory reporting, sanctions screening, trade surveillance, record retention, entitlement processing, and customer support. In regulated finance, the "edge" is rarely where the cost lives.

Identity is a perfect example. Institutions cannot transact with anonymous counterparties. They need to know who is on the other side, whether they are eligible to hold the asset, whether they are sanctioned, whether they meet investor-type rules, and whether transfers must be restricted by jurisdiction. You can build a beautiful token contract that enforces transfer restrictions, but then you need a robust identity and credentialing layer that is privacy-preserving, up-to-date, and legally defensible. You need processes for onboarding, offboarding, revocation, and remediation. You need to handle mergers, name changes, lost keys, and legal disputes. This is not a footnote; it is a large portion of the operating model.

Then there's the data problem. Many assets depend on off-chain facts: coupon rates, reference indices, rate resets, corporate actions, valuation marks, and legal notices. Smart contracts do not magically know these facts. They need trusted inputs, governance around those inputs, and audit trails. The more an asset relies on external data, the more the system resembles a carefully governed hybrid: on-chain state plus off-chain truth sources. Hybrid systems can be excellent, but they require discipline. Without discipline, you get a brittle machine whose failures are harder to debug because responsibility is split between code and process.

All of this costs money upfront. It also costs money ongoing: security reviews, smart-contract audits, key management, incident response, and governance upgrades. A mature ROI model treats these as first-class expenses, not rounding errors.

The adoption constraint nobody can code away: legacy inertia and "authoritative books"

Financial markets do not change by replacing one system; they change by agreeing on what counts as the authoritative record. In many markets, the authoritative books are held by central infrastructures—depositories, clearing houses, custodians—or by regulated roles like transfer agents and fund administrators. Tokenization can work with these roles (often the practical path), but it cannot pretend they don't exist. A tokenized system that lacks clear legal finality and recognized record-keeping is not a settlement system; it is an experiment.

This is why many credible tokenization deployments look incremental: a new token layer that is legally tied to existing record-keeping roles, or an interoperability approach that lets tokenized activity settle in recognized money while the legacy system still provides finality. Purists may complain that this is not "fully on-chain." Businesses will quietly celebrate because it is deployable. The business case improves when tokenization is used to reduce friction without demanding a Big Bang replacement of everything that already works.

There is also a human constraint: institutions are rightly allergic to operational outages. Post-trade plumbing is one of the most mission-critical parts of the economy. The tolerance for downtime is close to zero. That makes migration slow, staged, and full of parallel runs. Those parallel runs are expensive. A tokenization strategy that assumes immediate cutover is not ambitious; it is naïve.

What a real business case looks like: measurable, operational, and a little boring

A credible tokenization business case is not built on "the market will be trillions." It is built on specific operational metrics: reduced settlement fails, reduced breaks, reduced time-to-reconcile, fewer manual touchpoints per lifecycle event, faster collateral substitutions, lower margin buffers, shorter processing windows, improved auditability with lower compliance labor. These are boring metrics in the way that good infrastructure is boring: because it works.

It also looks like a clear operating model. Who is allowed to write to the ledger? Who validates? Who has upgrade rights? How are disputes handled? What happens when a key is lost? How are sanctions enforced? How does the system integrate into accounting and reporting? What's the incident response plan? A system that cannot answer these questions is not a cost saver; it is a risk generator.

And it includes an honest view of where savings will *not* appear. Tokenization does not eliminate credit risk; it can reduce exposure windows, but credit risk remains. It does not eliminate market risk; prices still move. It does not eliminate legal risk; it changes the surfaces where legal risk appears. It does not eliminate the need for governance; it makes governance more explicit. The business case is strongest when the value proposition is "do the same work with less duplication and fewer exceptions," not "rewrite finance."

A practical mental model: tokenization pays when it compresses the workflow

The simplest way to decide whether tokenization saves money is to draw the workflow and count handoffs. Not the marketing workflow—the real one. How many systems get updated? How many messages are sent? How many teams touch the process? How many times does the same data get re-keyed, reformatted, or reconciled? Each handoff is a cost center: it consumes time, it creates error, it requires controls, and it needs people.

Tokenization pays when it can collapse several of those handoffs into a single shared update—when messaging, reconciliation, and settlement become one coordinated operation rather than a relay race. It pays when it turns eligibility and compliance checks into machine-enforced gates that are auditable and consistent, rather than after-the-fact reviews that can be missed or disputed. It pays when collateral can be locked, monitored, and moved with fewer operational steps and clearer rules. It pays when it reduces the number of "versions of truth" that must be reconciled.

It does not pay when it adds a new layer without removing an old one. It does not pay when it demands heroic integration but delivers marginal operational change. It does not pay when it treats governance as an afterthought and then spends its savings on incident response. And it certainly does not pay when "tokenization" is just a branding exercise for a database that never needed to be distributed in the first place.

That's the uncomfortable conclusion: the business case for tokenization is real, but it is not universal. It's a scalpel, not a sledgehammer. Used precisely—against reconciliation, settlement latency, and collateral friction—it can cut out entire categories of waste that have been tolerated for decades. Used vaguely, it becomes a very expensive way to discover that finance already had computers.

7.2 Rewiring Wall Street: Bonds, Funds, and Private Markets

Wall Street has always been a strange blend of the futuristic and the antique. A trader can execute a billion-dollar bond switch in seconds, then spend the rest of the day waiting for confirmations, allocations, and settlement reports to line up across a small universe of systems. Tokenization targets that awkward gap: the place where finance stops being a market and becomes a paperwork factory.

The most useful way to think about tokenized capital markets is not as "assets on a blockchain," but as *instruments with a digital nervous system.* A traditional security is mostly inert. It sits in custody, it changes beneficial ownership through intermediaries, and its lifecycle events are pushed through a chain of agents and messages. A tokenized security can carry rules, permissions, and event logic in its own representation—so parts of the lifecycle become something closer to software execution than operational choreography.

That shift sounds abstract until you pick up familiar instruments-bonds, funds, and private-market interests-and ask an unromantic question: where does friction hide today, and what happens when the instrument itself can enforce some of the rules?

Tokenized bonds: turning a PDF term sheet into a living instrument

A bond is often described as simple: lend money, receive coupons, get principal back at maturity. The simplicity is real, but the machinery is not. Even plain-vanilla bonds require issuance workflows (bookbuilding, allocations, settlement), paying agents, registrars, custodians, and a surprisingly intricate calendar of ex-dates, record dates, payment dates, holidays, and cutoffs. In practice, a bond's "life" is coordinated by a web of institutions whose job is to keep the ledger straight and ensure entitlements go to the right owners.

Tokenized bonds push on a specific pressure point: the operational distance between *the legal promise* and *the operational execution*. In legacy rails, the legal promise is written in documentation, while the operational execution is performed by systems and agents interpreting that documentation. The bond does not "know" its own rules. It's more like a contract that humans and institutions continually reenact.

With a tokenized bond, some of that reenactment can be replaced by code that executes lifecycle events as state transitions. "State" here just means the bond's current truth: who owns it, whether it is encumbered, whether it is eligible to be transferred to a given buyer, what coupon is due, and what has already been paid. When those truths are updated on a shared ledger, and when the coupon logic is deterministic, the bond becomes easier to administer because fewer parties need to independently maintain (and reconcile) the same story.

The famous early proof-point for this idea was the World Bank's blockchain bond, often referred to as *bond-i*, issued in Australian dollars and managed across its lifecycle using distributed ledger technology. What made it notable wasn't a new kind of credit risk—it was the attempt to make the operational lifecycle more native to the digital representation: creation, allocation, transfer, and ongoing management living closer to one unified record rather than being stitched together across many ledgers. Even if the broader market didn't immediately migrate, the experiment highlighted an important design question: if the instrument can carry its own lifecycle logic, how much of the "bond bureaucracy" can be turned into routine computation?

The most seductive version of that story is the "smart bond": a bond that pays its own coupons automatically. That's not marketing poetry; it's a concrete pattern. If the bond exists as a token, and if the payment asset (tokenized cash, tokenized deposits, or another credible settlement instrument) also exists on compatible rails, then coupon payments can become an atomic event: entitlement calculation plus payment plus state update, all recorded with a single source of truth.

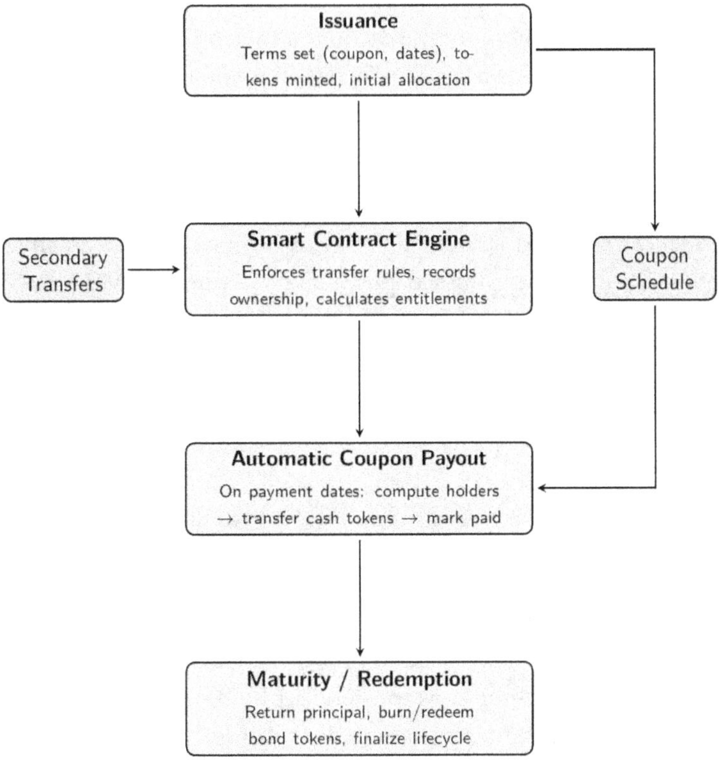

The diagram is clean; reality adds conditions. A bond can have rate resets, make-whole provisions, optional calls, tax gross-ups, and all the little clauses that lawyers love and systems engineers learn to fear. The smarter way to phrase the benefit is not "the bond pays itself," but *"the bond reduces the surface area for human interpretation."* Even partial automation-coupon calculation and entitlement logic, transfer restrictions, auditable event logs-can reduce the operational drag that makes fixed income feel heavier than it should.

The deeper improvement is how tokenization reshapes settlement and custody for bonds. In conventional markets, "ownership" is often layered: beneficial owners hold through custodians, who hold through depositories, who maintain the authoritative record. Tokenization doesn't have to abolish that hierarchy to be useful. It can make the hierarchy more transparent and programmable. A custodian can still exist as a regulated control point, but the asset's state transitions can be shared, time-stamped, and rule-bound in a way that reduces reconciliation and ambiguity.

And the unglamorous truth is that bond markets love anything that reduces failed settlements and operational breaks. A missed coupon is not just a missed payment—it's phone calls, investigations, client escalations, and reputational damage. If tokenization can make those events more deterministic and more auditable, the value shows up in fewer exceptions, fewer disputes, and less operational fire-fighting.

Tokenized funds: fewer gatekeepers, wider distribution, cleaner data

If bonds are about lifecycle events, funds are about *administration*. A mutual fund or money market fund is a machine that turns subscriptions and redemptions into pooled holdings, calculates a net asset value (NAV), tracks shareholder records, and processes flows with cutoffs and settlement conventions. The investment management may be sophisticated; the shareholder servicing is relentless.

Tokenized funds aim at an awkward bottleneck: fund shares are "digital" in the sense that they sit in systems, but they are often not digitally *portable*. Distribution can be expensive because it runs through platforms, transfer agents, intermediaries, and jurisdiction-specific plumbing. Even when everything is electronic, it can still be slow, fragmented, and surprisingly manual at the edges.

A high-profile example came when Franklin Templeton launched a tokenized share class representation for its on-chain money market fund offering, where shares are represented by a token (often referenced under the BENJI branding) and public blockchain infrastructure is used for transaction processing and share recordkeeping. The most interesting part of that story isn't the marketing headline;

it's the operational implication: you can represent "ownership" in a form that is easier to move between contexts and easier to integrate into digital experiences—while still running the fund under the familiar regulatory and administrative framework investors expect.

Fund tokenization is frequently misunderstood as "turning funds into crypto." It's closer to turning a fund share into something more like a modern digital instrument: a unit that can be held and transferred under rules, with ownership and transaction history represented in a shared, auditable way. The fund still needs a manager, an administrator, a custodian, and robust controls. The token is a new interface to those roles, not a replacement for them.

The practical business case for tokenized funds tends to cluster around three areas:

First, *distribution*. If the fund share can be represented as a token that is easy to integrate into apps and platforms, the marginal cost of reaching new investors can fall. That does not mean "anyone can buy it." It means the distribution channel can become more software-native, with eligibility and onboarding handled through digital identity rails rather than paper workflows and bespoke platform integrations.

Second, *transfer and settlement mechanics*. Fund subscriptions and redemptions are famously bound by cutoffs and processing cycles. Tokenization doesn't abolish those constraints—NAV calculation is still NAV calculation—but it can tighten the operational loop: a cleaner record of who owns what, fewer reconciliation points between distributor and transfer agent, and clearer audit trails.

Third, *data integrity*. Funds produce a constant stream of data: positions, flows, shareholder records, fees, and reports. Tokenization can allow portions of that data to be published (with the right permissions) in a way that downstream users can trust without endless file exchanges. That's why hybrid models are so revealing. BNY, for example, has described an on-chain approach that broadcasts trusted fund accounting data for a tokenized fund while the administrator and custodian roles remain central. That architecture is quietly pragmatic: keep the regulated operational core, but improve how information is shared and verified.

The misconception to avoid is that tokenized funds are automatically more liquid. Liquidity is not a property of a database; it's a property of a market. But tokenization *can* make liquidity easier to build by lowering frictions around transfer, recordkeeping, and access—especially for products that are currently sticky simply because the rails are sticky.

Private markets on-chain: the liquidity mirage, and the liquidity opportunity

Private markets are the promised land of tokenization pitches for a reason: they are enormous, illiquid, operationally heavy, and full of access constraints. Private equity, private credit, real estate partnerships, infrastructure funds—these instruments are valuable, but they are not built for easy transfer. Their friction is not an accident. It comes from information asymmetry, negotiated terms, investor eligibility rules, transfer restrictions, and the reality that many issuers prefer stable, long-term capital over a constantly churning investor base.

Tokenization can't repeal those economic preferences. What it can do is reduce the *mechanical* difficulty of creating controlled liquidity. And that difference matters.

A private-market interest today often behaves like a folder in a filing cabinet: subscription agreements, capital call notices, distribution statements, K-1s or tax documents, side letters, transfer approvals. The investor's "asset" is real, but it lives in documents and admin systems rather than in an instrument that can be cleanly moved. Secondary transfers-when they happen-can be slow and bespoke, requiring issuer consent, eligibility checks, and careful updating of the cap table and records.

Tokenization changes the cap table from a document to a state machine. Ownership becomes a controlled, permissioned ledger entry. Transfer restrictions can be encoded: only whitelisted investors can receive the token; transfers can require approvals; tokens can be frozen under legal orders; lockups can be enforced by code rather than by threatening language in contracts. None of this removes the need for governance, but it can turn governance into a repeatable workflow rather than a reinvention each time.

This is where "fractionalization" enters the conversation. Fractionalization means splitting ownership into smaller units—making it possible to buy a $1,000 slice of something that historically demanded $250,000 or $1 million. It's a powerful idea, and it is often oversold. Smaller units do not automatically create a buyer base. They can, however, create the *possibility* of a broader buyer base, which is not the same thing but is still valuable.

A better way to frame the opportunity is: tokenization can make private assets easier to package into products that investors can actually access—feeder funds, diversified vehicles, or controlled secondary markets—without the administrative overhead scaling linearly with the number of investors. The limitation is that private markets are constrained by information and governance as much as by mechanics. If the issuer does not want transferability, code won't change that. If disclosures are limited, a token doesn't magically protect buyers. The liquidity opportunity is real, but it tends to appear first in *managed* liquidity: periodic windows, approved venues, controlled participant sets, and clear rules.

The most interesting private-market effect may not be constant trading, but faster and cleaner *collateralization*. If a private credit position or fund interest can be represented as a controlled token with transparent encumbrance status, it becomes easier to pledge, monitor, and potentially finance. That creates liquidity in a subtler way: not by selling the asset, but by borrowing against it with less friction.

The settlement asset problem: the instrument can be digital, but the money has to be real

A tokenized bond that "settles" without a credible settlement asset is like a self-driving car without roads. You can admire the engineering, but you can't get anywhere.

In institutional markets, the gold standard settlement asset is central bank money—because it is final and free of commercial bank credit risk. That's why so much serious experimentation has focused on how tokenized assets could settle using central bank money, either directly or through interoperability with existing central bank payment systems. The Eurosystem ran exploratory work involving many market participants and real settlement in central bank money over a defined period, precisely because the question isn't theoretical: if you want scale in wholesale markets, participants need confidence in the settlement leg.

Switzerland's Project Helvetia explored similar territory, including pilots that used wholesale central bank digital currency concepts for settling tokenized assets. The message across these efforts is consistent: tokenization can modernize the asset leg, but the payment leg determines whether settlement is truly atomic, truly final, and therefore truly efficient.

This is also where the industry's near-term pragmatism shows up. Rather than waiting for a single perfect "tokenized money" solution, many architectures aim for interoperability: tokenized assets on a DLT platform that can settle via links to existing payment systems, or through tokenized deposits issued by commercial banks under familiar legal frameworks. It is less romantic than a clean-slate redesign, but it is how production systems usually evolve: by connecting the new machinery to the old power grid.

Democratization, with guardrails: access expands, but it does not become lawless

"Democratization" is a word that deserves suspicion in finance because it is often used as a halo rather than a plan. Expanding access to investments can be genuinely good—lower minimums, better distribution, more competition, more choice—but only if the product design respects investor protection, disclosure, suitability, and the fact that many restrictions exist because people can be harmed.

Tokenization changes the mechanics of access. It can reduce minimum investment sizes and distribution costs; it can make ownership easier to represent; it can make transfer rules easier to enforce. What it does not do is remove the need for legal categories like accredited investors, qualified purchasers, eligible counterparties, or jurisdiction-specific marketing rules. The realistic future is not a free-for-all marketplace; it is a more software-native marketplace where rules are enforced by design rather than by after-the-fact policing.

That distinction matters for institutions. A bank, asset manager, or market operator is rarely afraid of technology; it's afraid of controllability. Tokenization wins adoption when it offers *more* control where it's needed (eligibility, auditability, enforceable transfer rules) and *less* friction where it's wasteful (reconciliation, redundant record-keeping, manual lifecycle processing). That is how you get broader access without losing the guardrails that keep regulated finance upright.

What ``rewiring'' looks like in practice: less reinvention, more hybrid engineering

The most honest picture of tokenized capital markets is not a clean replacement of everything, but a rewiring: new rails running alongside old ones, gradually taking on more load as confidence grows. That's not a compromise; it's an engineering reality. Markets are not apps. You cannot ship a breaking change to the global settlement system and hope users update overnight.

Hybrid designs—where tokens represent instruments on-chain while core legal roles (custodians, administrators, transfer agents) remain central—often look less revolutionary on a conference stage. They tend to look more impressive in production, because they respect the two things markets care about most: operational resilience and legal finality.

And the punchline is that these hybrid models can still deliver the headline benefits that matter: faster and more reliable settlement patterns, tighter lifecycle automation, cleaner data distribution, and a path to controlled liquidity in places that are currently frozen by administrative overhead. The digital makeover isn't about making finance trendy. It's about making the same instruments feel less like paperwork and more like products.

7.3 ID Cards for Robots: Identity and Compliance-by-Design

The moment an institutional token project gets serious, the conversation stops being about throughput and starts being about a different kind of plumbing: *Who is allowed to touch this thing?* Not "who has the private key"—that's a crypto hobbyist's question—but who is legally permitted to buy, hold, transfer, pledge, or redeem the asset. In traditional markets, that question is answered with account openings, broker-dealer relationships, custodian controls, and a paper trail thick enough to stun a moose. On token rails, the asset can move with the ease of a text message. The compliance system has to move even faster, or the whole project collapses under its own risk committee.

Compliance is often described as a brake. In reality it's also a steering wheel. It decides which customers you can serve, which jurisdictions you can operate in, what kinds of liquidity you can offer, and how quickly you can move when regulators change the rules. Tokenization doesn't make compliance disappear; it makes it *explicit*. When assets become programmable and portable, the old informal assumptions—"our transfer agent will catch it," "our custodian won't allow it," "we'll reconcile it later"—turn into sharp failure modes. A token that can be transferred to anyone is not a product; it's a liability.

This is where "identity" enters, and it's easy to misunderstand. Identity on tokenized rails is not a giant dossier stapled to a wallet address. No serious institution wants a blockchain that publicly broadcasts home addresses and passport numbers. Identity, done properly, is closer to a set of *machine-checkable claims*: accredited investor status, KYC completed, not on sanctions lists, resident in permitted jurisdictions, within concentration limits, permitted to hold under a specific offering memorandum. The goal is not to reveal more data. The goal is to reveal *just enough*—and to do it in a way software can verify in milliseconds.

Why ``anonymous wallets" and ``regulated markets" don't mix

The simplest story of public blockchains is that anyone can create a wallet and transact. The simplest story of regulated finance is that you need to know who you're dealing with. These stories collide like weather fronts.

KYC ("Know Your Customer") is the baseline: identifying and verifying a customer. AML ("Anti-Money Laundering") adds monitoring and reporting. Sanctions screening adds hard constraints: certain people, entities, and jurisdictions are simply off-limits. Then come product-specific rules: accredited investor requirements, transfer restrictions for private placements, ownership limits for certain funds, restrictions on marketing and distribution, and the operational needs of tax reporting and shareholder communications. Each rule is manageable on its own; the problem is that traditional systems manage them through *intermediaries*. The broker, the custodian, the transfer agent, and the administrator become the enforcement points.

Tokenization changes the enforcement geometry. If a token can be transferred peer-to-peer, enforcement cannot rely solely on institutions sitting in the middle of every hop. Even if you prefer an intermediary model, tokens tend to travel through more diverse paths: new venues, new wallets, new forms of collateral movement, new cross-entity workflows. The enforcement has to be more *portable* too.

That portability is what "compliance-by-design" really means. It means the compliance checks are not merely performed at onboarding and then forgotten. They are continuously enforceable at the moment of transfer, pledge, redemption, or other lifecycle events. If the rules change, the system can update its enforcement logic without rewriting every participant's internal procedures from scratch.

ID cards for robots: verifiable credentials in plain English

A useful mental model is the one used at airports. You don't hand the gate agent your entire life story. You hand over a credential that answers a narrow question: *Are you allowed to board this plane?* The credential might be tied to a deeper identity system, but the interaction is minimal and purpose-specific.

A *verifiable credential* is the digital version of that boarding pass. It's a standardized way for an issuer (say, a bank, an identity provider, or a regulated onboarding firm) to sign a claim about a user. The user holds that credential—often in a digital wallet—and can present it to a verifier to prove something without exposing unnecessary details.

The crucial word is "verifiable." The verifier can cryptographically check that the credential was issued by a trusted issuer and hasn't been tampered with. And because real life changes, credentials can also be revoked or marked invalid through status mechanisms—so a "not sanctioned" credential isn't a forever-pass if sanctions lists update next week.

This is where tokenization becomes less like "finance on a blockchain" and more like "finance with modern identity infrastructure." A smart contract doesn't need your passport scan. It needs a yes/no answer to questions like: is this wallet controlled by an eligible investor? is the investor in an allowed jurisdiction? is the investor currently sanctioned? has the investor passed KYC for this product? Those answers can be delivered as cryptographic proofs or attestations.

The practical payoff is speed and consistency. When eligibility is machine-checkable, distribution can become less artisanal. A tokenized fund share doesn't have to be distributed through a handful of platforms because onboarding is too expensive elsewhere; it can be distributed through many channels that all rely on the same credentialing logic. That's not deregulation. It's better enforcement with less friction.

Privacy isn't optional: institutions need discretion, not exposure

There's a hidden reason institutions like the idea of credential-based identity: it can be *more private* than the current system. That sounds backwards until you consider how compliance works today. A customer submits a large package of documents to multiple institutions. Each institution stores them, replicates them, and becomes a new breach risk. Your identity is "protected" by being copied into a dozen databases, each guarded by its own policies and staffing levels. It's an odd kind of safety.

A credential model can reduce data sprawl. Instead of sending raw documents everywhere, you can send signed attestations that are scoped to purpose. "KYC completed by a regulated institution" can be proven without sending the entire KYC file. "Accredited investor under a particular rule" can be proven without disclosing net worth.

The verifier gets enough confidence to allow the transaction, while the user reveals less.

Of course, privacy is not a free lunch. Someone still has to do the underlying verification. Regulators still need audit trails. Firms still need to retain records and explain decisions. The point is not to hide from oversight; it is to reduce unnecessary exposure and replication while preserving accountability.

Permissioned tokens: when the asset itself enforces the rules

A quiet reality of institutional tokenization is that many assets will be *permissioned*. That word makes some people flinch because it sounds like "blockchain with gatekeepers." In finance, gatekeeping is often the product. The question is whether gatekeeping is enforced by brittle manual processes or by auditable, consistent code.

Permissioned token standards and patterns typically split compliance into two layers:

- An *identity registry* that maps wallet addresses (or accounts) to verified identities or credential statuses.
- A *compliance module* that checks rules at the moment of transfer or other token actions.

One widely discussed pattern in the tokenization ecosystem is the ERC-3643 family of approaches, which is designed specifically for permissioned securities. The details vary by implementation, but the philosophy is consistent: transfers are not purely "whoever has the key wins." They are conditional on identity verification and compliance rules.

Why does this matter for the business case? Because it converts compliance from a series of external controls into a property of the instrument's operation. If a private-market token cannot be transferred to an unapproved buyer, you don't need to rely solely on downstream detection. The prevention happens at the gate. If a regulator requires immediate freezing of assets associated with a sanctioned entity, a permissioned design can support freezing and forced transfers under controlled governance. Those powers are controversial in retail crypto culture; in regulated finance they're closer to standard operating procedure, just implemented through intermediaries today rather than through an explicit token governance model.

There is, however, a hard trade-off: the more compliance logic you embed, the more you must treat the token system like mission-critical infrastructure. Bugs in compliance code are not "minor defects." They are potential regulatory breaches. This forces maturity: formal testing, audits, change management, incident response, and clear accountability for who can update rules. Compliance-by-design is not a shortcut around governance; it is a reason governance must be stronger.

The Travel Rule and the unavoidable reality of cross-border compliance

Crypto's early narrative treated cross-border transferability as a kind of natural right. Regulated finance treats cross-border flows as a compliance event. The "Travel Rule" in AML/CFT regimes requires that certain information about originators and beneficiaries "travels" with a transaction between service providers. It's an attempt to make digital value transfers less opaque to law enforcement and regulators.

Tokenization collides with the Travel Rule in a specific way: if assets can move quickly across entities and jurisdictions, compliance information has to move with them, reliably and in standardized formats. The painful part is that implementation has been uneven across jurisdictions and institutions. Even where rules exist, supervision and enforcement can lag, and firms are left navigating a patchwork of expectations. That patchwork is not a theoretical inconvenience; it becomes a design constraint.

This is one of the strongest arguments for building compliance into the rails rather than layering it on later. If firms attempt to "bolt on" Travel Rule processes to token transfers after the fact, they often recreate the same manual workflows tokenization was supposed to reduce—only now with faster-moving assets and more failure modes. If, instead, identity attestations and counterparty checks are part of the transaction workflow, compliance becomes less of an external interruption and more of a native step.

Still, the industry should be honest: tokenization will not make cross-border compliance easy. It will make it *more automated* for the flows that are already permitted, while making it *more enforceable* for flows that are not. That is progress, but it's not frictionless.

The ``frozen funds" problem: operational contingencies are not edge cases

Every tokenization pitch loves the happy path: investor buys token, token settles instantly, everyone smiles. Real markets are built for unhappy paths. Keys get lost. Trustees die. Investors get hacked. Courts issue orders. Regulators impose freezes. Issuers make mistakes. Even in traditional markets, these events require messy interventions: stop transfers, reverse transfers, correct records, reissue certificates (in older systems), and coordinate among custodians and agents.

A tokenized system that cannot handle these contingencies is not ready for institutional scale. And a system that *can* handle them usually needs explicit administrative functions: pausing transfers, freezing accounts, forced transfers under legal authority, recovery mechanisms, and auditable logs of who did what and why.

This is where compliance-by-design stops sounding like a philosophy and starts sounding like an operating manual. The token's rules are not only about preventing bad transfers; they are about making necessary interventions precise and accountable. It's better to have a clearly governed freeze function than to rely on informal "we'll call the custodian" procedures that differ by firm and fail under stress.

Of course, these controls raise a legitimate concern: they create power. Who holds it? Under what legal authority? With what oversight? How are governance keys secured? How are actions logged and reviewed? This is not a reason to avoid tokenization; it is a reason to treat token governance like financial infrastructure governance, not like a startup feature flag.

Compliance as a product feature: why ``boring" controls unlock real distribution

There's a counterintuitive point that keeps showing up in successful institutional designs: compliance is not just a constraint. It can be a *feature*. If an asset can prove, at the moment of transfer, that it is being held by eligible investors and that restrictions are being enforced, it becomes easier to distribute through more channels. Platforms and intermediaries become more willing to carry it. Risk committees relax. Regulators become more comfortable with experimentation when guardrails are explicit.

This is especially important for the "digital makeover" of funds and private-market products. Their distribution has always been limited by onboarding cost and compliance overhead. If credentials and permissioning reduce the marginal cost of adding a new distributor or a new qualified investor segment, that's not a theoretical efficiency. That's revenue expansion through lower friction.

The same logic applies to secondary markets for private assets. Secondary liquidity is often throttled not by lack of interest, but by compliance friction: transfer approvals, investor checks, paperwork, and record updates. If transfers can be made compliant by default—enforced in the token's rules and supported by verifiable credentials—controlled liquidity becomes easier to operationalize. Not effortless. Just finally *repeatable*.

The uncomfortable truth: automated compliance still needs humans

It's tempting to imagine compliance-by-design as a machine that runs without people. The better analogy is autopilot. Autopilot can fly the plane for long stretches, but it requires a certified crew, rigorous maintenance, and clear procedures for when the unexpected happens. Automated compliance reduces routine manual checks and increases consistency, but it does not eliminate judgment, escalation, and oversight.

Someone must decide what "eligible" means in a specific product. Someone must select trusted credential issuers. Someone must define revocation policies and monitor exceptions. Someone must respond to false positives and false negatives. Someone must handle disputes: the investor who insists they are accredited; the entity flagged by a sanctions screening error; the investor whose credential expired because an issuer had an outage. Code can enforce rules, but it cannot explain them to regulators or clients. Humans still do that part.

This is why the strongest institutional tokenization efforts treat compliance as an engineering discipline: product requirements translated into rules, rules translated into code and credential schemas, code backed by governance, monitoring, auditability, and the ability to adapt when laws and regulations change. The "ID cards for robots" are only useful if the robots are operating in a city with real laws—and if the city can update the traffic rules without shutting down every road.

When this works, the payoff is not merely fewer forms. It's a new kind of market ergonomics: instruments that can move faster *because* the guardrails are tighter, not looser. That's the paradox at the heart of institutional tokenization: speed comes from control, and control comes from identity that machines can understand.

7.4 Speaking the Same Language: Standardization and Coordination

The fastest way to kill a tokenized-market project is to make it "work" perfectly—inside your building. A demo that settles instantly between two desks on the same platform can look like the future. Then someone asks a rude question: *Can it settle with someone else's platform?* Can your tokenized cash pay for their tokenized bond? Can your identity rules be recognized by a counterparty who doesn't share your onboarding vendor? Can your definitions of "owner," "pledge," and "finality" survive contact with an auditor, a regulator, and a cross-border custodian?

Finance is not a product; it's a coordination game played at industrial scale. The real obstacle isn't cryptography. It's agreement. Tokenization doesn't just add a new technology stack—it adds new ways to disagree. Different chains, different token standards, different wallet models, different legal wrappers, different compliance regimes, different message formats, different settlement assets. Without shared language and shared interfaces, the industry risks building a glittering archipelago of "digital islands," each modern in isolation and useless in combination.

The irony is that capital markets already learned this lesson the hard way. The modern system is stitched together by standards so old they feel like weather: identifiers, messaging formats, legal definitions, market practice conventions, payment rails, settlement cycles. People complain about them until they imagine the alternative: every institution inventing its own vocabulary and demanding everyone else translate. That's not innovation. That's a tax on every trade.

Interoperability isn't a feature; it's the whole point

Interoperability is a word that gets tossed around like a nice-to-have, as if it were Bluetooth on a new gadget. In markets, interoperability is the difference between infrastructure and a science project.

A token's value comes from the ability to be used across organizational boundaries. A bond that can be bought only by people on the issuer's preferred platform is not a market instrument; it's a private database record. A tokenized fund share that cannot be held at a custodian because the custodian's systems don't recognize the token's transfer rules is not "digital." It's stranded. The entire economic promise of tokenization-more fluid collateral, faster settlement, broader distribution, controlled secondary liquidity-depends on assets and money meeting each other across systems.

Regulators have been blunt about this: tokenization can amplify operational and legal complexity, and fragmentation can turn into a systemic risk rather than a harmless inconvenience. Fragmentation in a consumer app ecosystem is annoying. Fragmentation in settlement infrastructure can create settlement fails, liquidity traps, and disputes about which record is authoritative. When "the ledger" splinters, the industry doesn't get multiple truths; it gets multiple lawsuits.

The hidden costs of ``digital islands"

A digital island is not merely a platform that doesn't talk to others. It is a platform that forces everyone else to build bespoke bridges. Bespoke bridges are expensive, brittle, and politically fragile because every bridge becomes a negotiation about data, liability, governance, and downtime.

The first cost is integration. Each new platform demands new connectivity, new security reviews, new operational procedures, new training, new support. Integration costs don't scale linearly; they explode combinatorially. If there are N islands and each needs bespoke connections to the others, the number of bridges grows roughly like $N(N-1)/2$. That's a grim equation to put in front of a CFO.

The second cost is operational risk. Each bridge is a place where messages can be misinterpreted and where state can drift. A token might say "settled" on one platform while the payment leg is pending on another. One platform might treat a corporate action as final while another treats it as reversible. If the bridges rely on humans, you get exceptions and delays. If they rely on code, you get software risk and upgrade coordination problems.

The third cost is liquidity fragmentation. Liquidity is a social phenomenon: it appears when participants believe they can trade quickly at predictable prices. Fragmentation splits order flow. If some holders are locked into one island and others into another, the market becomes thinner in both places. The token may technically be transferable, but practically illiquid because participants can't easily meet in the same venue with the same settlement expectations.

And then there's the cost that hurts the most: lost optionality. A truly interoperable asset can be used as collateral, traded, pledged, redeemed, and integrated into multiple workflows. A stranded asset can do one thing in one place. Tokenization's economic case leans heavily on optionality-so fragmentation strikes at the heart of the story.

Standards are the unglamorous technology that makes everything else possible

If tokenization is the flashy new building, standards are plumbing and electrical codes. Nobody tours a city to admire its codes. But every city that ignores them eventually experiences a fire.

Standards show up in three interlocking layers:

Data standards: how fields are named and structured. What exactly is "settlement date"? Is it UTC? Local market time? Does it respect holidays? How do you represent partial fills, cancel-replace events, or a rate reset? When two systems disagree about the meaning of a field, automation turns into confusion at machine speed.

Process standards: how lifecycle events are modeled. A bond coupon, a fund subscription, a collateral pledge-these aren't just data points. They are sequences of events with conditions, deadlines, and consequences. If one firm represents a lifecycle as a single event and another represents it as five steps, shared automation breaks down.

Legal and semantic standards: what words mean in enforceable terms. "Ownership," "beneficial owner," "final settlement," "pledge," "novation," "transfer restriction." These aren't just labels; they are anchors for rights and obligations. Tokenization introduces new mechanics, but the legal vocabulary still determines what happens in a dispute.

Token standards in the blockchain world-ERC patterns and similar families-are attempts to create shared interfaces for asset behavior. Traditional finance has its own long-running standardization efforts, and the tokenization wave is forcing these worlds to make eye contact. The good news is that finance already knows how to do coordination when it must. The bad news is that it is rarely fast, never painless, and always political.

ISDA CDM: the underrated bridge between lawyers and machines

One of the most consequential standardization efforts in modern markets isn't a blockchain standard at all. It's the ISDA Common Domain Model (CDM), an attempt to create a shared, machine-readable model of financial products and their lifecycle events-especially in derivatives, where post-trade complexity is a daily reality.

The CDM matters because tokenization without shared lifecycle definitions is just digitized confusion. A "smart contract" that automates a lifecycle event still needs a canonical definition of that event.

What exactly is a coupon payment event? What is the record date? What is the entitlement calculation? What happens when a payment fails? What happens in a rate reset? What happens in a corporate action? If every institution encodes its own interpretation, automation doesn't scale; it multiplies bespoke logic.

The CDM is not a magic wand, and it won't replace firm-specific policies overnight. But its direction is important: it treats lifecycle events as objects that can be defined, shared, and executed consistently. Tokenization has a natural affinity for this approach. The more market participants can agree on lifecycle semantics, the more feasible it becomes to build interoperable smart contracts and interoperable post-trade processes.

Another benefit is psychological: shared models reduce the fear that tokenization is a leap into a proprietary unknown. When governance committees can point to an industry model, they are more willing to approve automation. Standards don't just align systems; they align risk appetites.

ISO language: boring vocabulary, sharp edges

If you've ever watched two engineers argue about the meaning of a term, you've seen why vocabulary matters. In finance, the stakes are higher because words become contracts and contracts become court cases.

ISO standards for blockchain and distributed ledger terminology-such as ISO 22739:2024-sound like the driest imaginable reading. They are also a surprisingly important stabilizer. When an industry is hot, everyone coins new phrases. When an industry becomes infrastructure, it needs definitions that survive hype cycles.

The impact is subtle but real. When regulators, auditors, banks, and technology vendors use the same vocabulary, discussions become less about translation and more about substance. That's not academic pedantry; it prevents expensive misunderstandings. A project can die simply because one stakeholder hears "token" and thinks "unregulated crypto asset," while another hears "token" and means "digitally represented security under existing law." Shared vocabulary doesn't resolve disagreement, but it makes disagreement visible—and therefore manageable.

ISO 20022: a reminder that big coordination can actually happen

Financial standardization is often mocked as glacial. Then ISO 20022 happened.

ISO 20022 is a global standard for financial messaging-how payments and related information are structured and communicated. Its migration has been a multi-year, high-stakes coordination effort across banks, market infrastructures, and vendors. The detail is messy, but the lesson is clean: the industry *can* execute large-scale cutovers when the incentives are aligned and when there is a credible governance process for timelines, coexistence periods, fallbacks, and exceptions.

Swift's management of ISO 20022 coexistence and its eventual end date for certain message types is a particularly useful analogy for tokenized markets. Tokenization will not be adopted via a single overnight switch. It will be adopted through coexistence: old rails and new rails running in parallel, translation layers bridging formats, contingency procedures for when messages can't be translated cleanly, and strong incentives to move because the old formats become less supported.

That is exactly the kind of boring, disciplined coordination tokenization needs. Without it, every institution will claim to support interoperability while quietly demanding everyone integrate to their preferred stack. Interoperability won't be declared; it will be scheduled, tested, governed, and—eventually—enforced by the practical reality that the market moves to where the friction is lowest.

Settlement assets: speaking the same language about ``money" is the hardest part

It is relatively easy to standardize a token interface for an asset. It is harder to standardize what counts as settlement money, because money is not just technology—it is trust and law.

In wholesale markets, participants care intensely about settlement in central bank money or close equivalents. That's why central banks and public-sector actors have focused on architectures that integrate or interoperate with central bank settlement systems, rather than imagining that market participants will simply accept any token as "cash." This isn't conservatism for its own sake. In a crisis, the definition of final money becomes existential.

The tokenization ecosystem is full of competing settlement instruments: stablecoins, tokenized deposits, central bank digital money concepts, and hybrid models that use existing payment rails. Each has different risk profiles, legal treatment, and operational implications. Without coordination, the market could splinter into cash silos: a tokenized bond that settles only against a particular bank's deposit token; a fund token that settles only against a particular stablecoin; collateral that is recognized only within a closed network. That is how you recreate the inefficiencies tokenization promised to dissolve.

Standardization here is not only technical. It includes legal frameworks, prudential treatment, and operational conventions: redemption mechanics, cutoffs, liquidity management, disclosures, and the handling of failures. Money is the shared language of markets; if tokenized money becomes a dialect spoken only on one island, interoperability loses much of its value.

Coordination as governance: who decides, who upgrades, who is liable

Even with perfect standards, systems still evolve. Rules change. Bugs are discovered. Security threats emerge. Regulations update. Tokenization adds a specific governance challenge: smart contracts and protocols can be upgraded, but upgrades require coordination among parties who do not share a boss.

Traditional market infrastructure handles change through governance bodies, rulebooks, and controlled release cycles. Tokenization needs the same seriousness, even when the underlying technology is open-source. If the industry wants interoperable rails, it must also want interoperable governance: predictable processes for proposing changes, testing, certifying, and rolling out upgrades, with clear accountability when something goes wrong.

Liability is the quiet killer here. When a message format is ambiguous and a payment goes missing, who is responsible? When a compliance rule encoded in a token blocks a legitimate transfer, who compensates the harmed party? When a bridge between systems fails and creates inconsistent state, which record is authoritative? The more "programmable" the system becomes, the more important it is to define responsibility for both code and operations.

This is where incumbents—central securities depositories, clearing houses, major custodians, payment system operators—often have an advantage. They already have governance structures and legal relationships. Tokenization doesn't have to replace them to achieve interoperability; it can use them as anchors, provided the new rails do not become proprietary moats.

The converging path: fewer islands, more shared rails

A fragmented digital finance world is not just inefficient; it is dangerous. It invites operational fragility, liquidity traps, and disputes about finality at the worst possible moments. The antidote is not one chain to rule them all. The antidote is agreement on interfaces, semantics, and governance—enough shared language that different implementations can still participate in the same markets.

Standards are often treated as the end of innovation. In finance, they are the beginning of scale. Tokenization becomes economically meaningful when a token issued in one place can be used in many places: traded, pledged, settled, reported, audited, and governed with predictable behavior. That requires shared definitions of events, shared representations of data, shared expectations about settlement money, and shared procedures for change.

The industry already knows how to coordinate when it must. The challenge is to do it before tokenization hardens into incompatible local optimizations—before the islands get too comfortable, and the bridges become too expensive to build.

www.ingramcontent.com/pod-product-compliance
Lightning Source LLC
LaVergne TN
LVHW020233210926
848481LV00016B/1793
9798896653103